PIERRE DUHEM

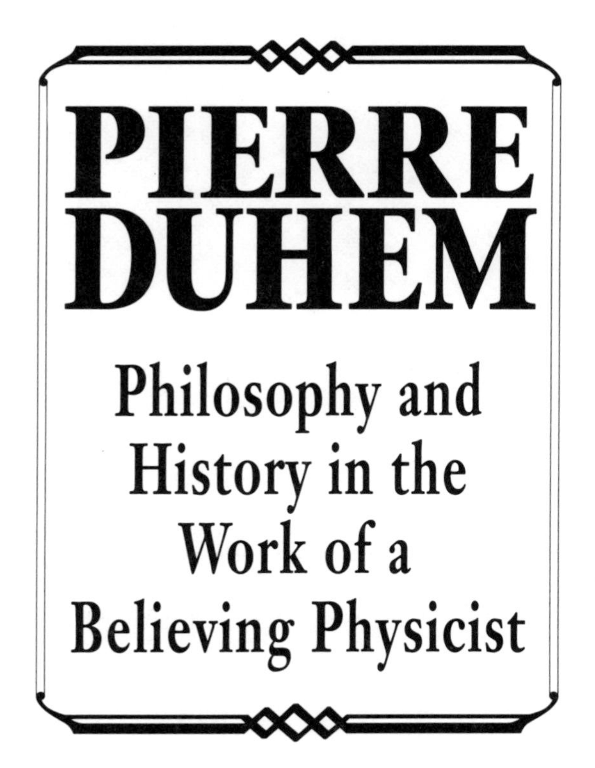

PIERRE DUHEM

Philosophy and History in the Work of a Believing Physicist

R. N. D. Martin

La Salle, Illinois

First printing 1991

Printed and bound in the United States of America.

Library of Congress Cataloging-in-Publication Data

Martin, R. N. D. (R. Niall D.)
Pierre Duhem: philosophy and history in the work of a believing physicist / R. N. D. Martin
p. cm.
Includes bibliographical references and index.
ISBN 0-8126-9159-8. — ISBN 0-8126-9160-1 (pbk.)
1. Duhem, Pierre Maurice Marie, 1861-1916. 2. Physics—History. 3. Science—History. 4. Physicists—France—Biography. I. Title.
QC16.D925M37 1991
530' . 092—dc20 91-26643
[B] CIP

Contents

Preface; By Way of Thanks

This essay began back in 1967 as a doctoral project at University College London, taken up at the suggestion of Dr. (now Professor) Larry Laudan. At that time Duhem was much talked about among philosophers and historians of science, but there was little serious systematic study of him—the work of Dr. Donald Miller was the exception. Given the enormous bulk of Duhem's writings there was no prospect of remedying this situation single-handedly but the narrower focus of Duhem's writings on the philosophy of science seemed more manageable. Perhaps I could mount an extended examination of his philosophical views? But that project turned out to be slippery and elusive; it was hard to make much sense of the debates surrounding Duhem and I was held up by illness and the need to earn some kind of a living. In 1974, however, the relation between the history and the philosophy of science was high on the agenda, in large measure because of the work of the late Professor Imre Lakatos. Since Duhem had worked in both areas, he was a prime candidate for a case study and I took the hint. There was, however, a prior problem I had been conscious of from the beginning; as a historian Duhem was known for his work on mediaeval science, but all of the work that earned him that reputation had come late in his career. I needed to know why before I could offer a credible analysis of the relation of Duhem's historical work to his philosophical work. I quickly found that mediaeval science had come to Duhem as a total surprise, a surprise, moreover, that thanks to Duhem's publication habits I could date to within a month. But in a Catholic the surprise was itself surprising, in view of the ecclesiastically-sponsored revival of Scholastic thought going on all around him: Duhem should not have, one would have thought, been surprised. I needed to understand better the implications of Duhem's Catholicism in its historical context.

Dr. Jerry Ravetz has been my mentor in all my subsequent explorations. Though a Protestant I did bring to the religious issues involved in this question an amateur interest in theology gained long before in the Student Christian Movement, but there was much to learn as I probed an unfamiliar field. It was the late Dr. Alec Vidler who first alerted me to Duhem's lifelong friendship, mentioned in none of the biographies, with the philosopher and theologian Maurice Blondel. The late Dr. Charles Duthie gave me valuable guidance on the Pascal literature. Dr. Donald Miller supplied copies of correspondence to Duhem when it was on loan to him. (It has since been deposited with the Académie des Sciences, Paris) and Alan and Françoise Prentice helped to transcribe some of it. I am particularly grateful to Mme G. Mosseray of the Archives Blondel, Louvain-la-Neuve, Belgium, for copies of Duhem's letters to Blondel and for commenting on my transcription of them. For various other kinds of help I thank Professor D. W. Hardy, Dr. Émile Poulat, Mr. Duncan McGibbon, Professor Harry Paul, and the late Dr. Charles Schmitt. In addition to the British Library, Reference Division, I am grateful to have used the Bibliothèque Nationale, the Bodleian Library, Oxford, the Catholic Central Library, the Science Museum Library, and the libraries of Edinburgh University, Heythrop College London, University College London, and the University of London. I also thank my family and friends for bearing with me all these years when it seemed that I might never finish.

In thanking all these I am only conscious of the incompleteness of the result. It is no longer true that we lack serious systematic studies of Duhem's work; we have Professor Jaki's massive biography with many invaluable references to out-of-the-way literature, Roberto Maiocchi's equally massive Italian work and now Anastasios Brenner's French thesis. But there is still so much to be done. Even in the religio-political area there is so much that remains obscure, while we still lack a systematic evaluation of Duhem's rôle in late nineteenth-century and early twentieth-century physics.

I

Introduction

1. Life and Activity

Duhem was not a minor figure; minor, neither, were his contributions to any of the fields he touched. He knew it, his friends knew it, and when they obstructed his career his enemies knew it too. When the French Académie des Sciences finally granted him non-resident membership in 1913, he had long since acquired honorary membership of several foreign academies. A nearly complete list of his writings has well on 400 items,[1] several of them massive multi-volume works, with none of the multi-author papers that feature in the publication lists of some of the more prolific modern scientists. Alike in physics, philosophy of science, and history of science, his influence is still felt, several of his works are currently in print, and translations of his works from the French continue to appear.

Pierre Maurice Marie Duhem was born in Paris in 1861, the son of a salesman of Flemish descent, and of a mother with roots in the south of France. Two of the family of four died of diphtheria in childhood; only Pierre and his sister Marie grew to maturity. This deeply Catholic family sent their son to Catholic schools: to the École St. Roch and the Collège Stanislas where, as he later reported, a teacher called Jules Moutier got him to love the theories of physics.[2] To study that subject after a bout of the ill-health that was to dog him all his life, he entered the élite École Normale Supérieure in 1882, first in the entry competition for his year.[3] In 1885, at the end of the normal three years for *license* and *agrégation*, he had completed a doctoral thesis as well, which, *cause célèbre*, was turned down by a jury of physicists headed by the future Nobel prize winner Gabriel Lippmann. Duhem, however, soon produced another, which the *mathematicians* thought worthy of a doctorate, while he, meantime, continued at the École Normale for the next two years with the select group of *agrégés-préparateurs*. In 1887, he became a highly successful *maître des conférences* (lecturer) at Lille, where he was to stay until a quarrel over the conduct of examinations forced a transfer to Rennes. The following year saw another transfer, this time to a professorship at Bordeaux, where, denied the prestige of a Paris chair, he was to remain until his death by a heart attack in 1916. His marriage to Adèle Chayet resulted in one daughter Hélène. Pregnant for a second time, Adèle died in childbirth.

The scope of his work is easily indicated: mathematical physics, philosophy of science, and history of science. The physics, the area of his initial training and main subsequent activity, far outstrips the other two in bulk, while the history comes second. In physics, he worked extensively on the theory of heat and its implications for other areas of physics, on fluid flow, and on electricity and magnetism. In philosophy of science there is one seminal monograph and a number of shorter articles. In the history of science there are a number of schematic works on the history of modern physics, and voluminous investigations into mediaeval and early modern physics.

2. The Problem of Duhem's Standing

None of this was of minor importance: the work on the theory of heat continues to influence the presentation of that subject by physical chemists at least; that on hydrodynamics has been reprinted by the French Ministère de l'Air;[4] that on electricity and magnetism was taken seriously by A. O'Rahilly in the early 1930s.[5] All this, and much else, was underpinned by a formidable mathematical expertise that drew the admiration of no less a judge than his friend Jacques Hadamard.[6] Outside physics, Duhem's work has also made its mark. He never considered himself a philosopher, but his *Théorie Physique* is at the centre of continuing debates in the philosophy of science.[7] Not primarily a historian, his major work on the history of mediaeval science has proved seminal for the subsequent development of the subject: the three volumes of the *Études sur Léonard de Vinci* are frequently cited by historians and were reprinted in the post-war period; the remaining volumes of the massive *Système du Monde* (unfinished at Duhem's death) were finally published in the 1950s, all ten volumes are currently in print, and a translation of important parts of the later volumes by Roger Ariew[8] has recently been published. Of the shorter historical works, there are German and English translations of the *Évolution de la Mécanique;*[9] *Le Mixte et la Combinaison Chimique* has recently been reprinted, as has also *To Save the Phenomena*; there is a translation of the latter into English,[10] and it too is the subject of continuous debate. Duhem was no minor figure.

Not a minor figure, but a major one? It is here that the reservations begin. Though the very violence with which they are sometimes urged may perhaps serve to undermine them, reservations nevertheless there are, and Duhem, after all, does not attract anything like the weight of citations of the likes of Henri Poincaré and Maurice Blondel, to name only two of his better-known contemporaries. Reservations concern every area of his activity, but it is well to start with the physics that was Duhem's first love.

In France, a major physicist would normally expect to see his career crowned by a Paris chair and membership of the Académie des Sciences: Duhem never got his Paris chair, and *non-resident* membership of the Académie came late in his career, in 1913, much later than all his friends had predicted, and when he had only three years left to live. He did receive recognition from several foreign academies, but these did not include the Royal Society of London, and he did not get the Nobel Prize. Not a participant in the Solvay conferences so much discussed by historians of modern physics, he played no part in the development of the particle-atomistic approach that now dominates all teaching and research in the subject. Though there are other reasons, to be discussed below, for this state of affairs, one obvious factor can be mentioned now: his opposition to the atomic theories that were to win the day in the early years of this century. From a point of view common in the 1880s when he came to maturity, he sought in an elaborated, formalized version of the theory of heat the basis for a comprehensive approach to the whole of physics that made no appeal to hidden molecular motions: his physics was to be a coherent *general thermodynamics* offering a general, but precise, mathematical description of the experienced world, of the phenomena revealed by detailed experiment. While in the early 1890s this approach seemed to be carrying all before it, in the following decade the tide turned and Duhem was sometimes lucky to sell a single copy of one of his books in a whole year. Similarly, few supported his long campaign against the electromagnetic theory of Maxwell in favour of that of Helmholtz.

One would hardly expect to find similar doubts raised about philosophical work that attracts at least a dozen references a year, but doubts there are. Duhem's *Théorie Physique* may indeed have recently been reprinted,[11] and it may indeed be the focus of intensive discussion, but that discussion largely boils down to a close critical examination of certain famous 'sceptical' arguments in part of the work, by writers who show little interest in its overall structure and tendency.[12] The reason for this will be discussed below, but *prima facie* it seems to lie in a perceived tension in his work: though the reference point for nearly all of it seems to be the phenomena as they are experimentally determined, underpinning arguments hostile to any notion of truth beyond the appearances, any form of so-called realism, an approach customarily called 'instrumentalist' or 'positivistic', Duhem wishes at the same time to sanction just such a notion of a truth behind the appearances. In the light of this apparent inconsistency Duhem is liable to seem a second-rank figure who happened to devise one or two billiant sceptical arguments. Alternatively, it will be hard to dismiss the suspicion that Duhem is keeping something back, that he is playing dishonest games with readers who could be well advised to take care before accepting any of his claims. Duhem's own extended discussion of his Catholicism in his 'Physique de Croyant' has done little to allay such doubts.

With Duhem's historical work, the story is repeated on a much larger scale, with additional complications. The apparent claim of the *Études* that certain theologically-motivated condemnations in Paris in 1277 marked the birth of modern science continues to strike many modern readers as simplistic at least, if not bizarre. To such readers, the work may seem dominated by a crude search for precursors for later ideas that does scant justice to the complexities of intellectual history and the differing contexts in which apparently similar ideas are expressed, a search sometimes dubbed 'precursoritis'. Hardly more persuasive has been Duhem's treatment of the Galileo affair in *To Save the Phenomena*, where the apparent attempt to defend the ecclesiastical personalities involved has seemed to many no more than misguided defence of the indefensible. both in Duhem's historical work and in his philosophical, many have found it natural to compare his work with that deriving from the revival of the teachings of St. Thomas Aquinas then coming to dominate Catholic thought.[13]

Quite apart from such reactions to the details of Duhem's work, there is the persistent feeling that Duhem is always overplaying his hand, defending the indefensible long after any reasonable doubt has been decisively removed. Duhem was a combative figure who attacked the great and the powerful without fear or favour. The rejected thesis referred to above, which Duhem promptly published, was partly directed at the views of the physicist, chemist, historian and republican politician Marcellin Berthelot, and Duhem was to return repeatedly to the charge.[14] Berthelot was in a position to obstruct Duhem's career, and it is usually thought that he did just that. Berthelot was not the only well-placed target Duhem took on. All this fierce polemic inevitably strikes many of his readers as odd if not sick, and make it harder for them to take him seriously.

3. The Ariadne's Thread

No minor figure, if the doubts could be removed Duhem might be accepted as a major one: the problem is how to deal with these doubts obstructing a proper view of his overall standing. They all seem to concern, one way or another, his agenda, the nature of the programme underlying his various interests and relating them to each other. By 'agenda' I include, not only the overt objectives Duhem stated explicitly, but also what is sometimes called the 'hidden agenda', the things not said, deliberately or otherwise, but which would follow if the explicit theses were accepted, or, even if they did not strictly follow, would be hard to resist. An example of what is involved here is provided by Abel Rey's response to Duhem's views. As will be explained below, Duhem's account of physical theory made it as useless for defending the Catholic faith as for attacking it. To Rey, nonetheless, Duhem's denial, in his *Théorie Physique*, of claims that physics could reach a reality behind the appearances implied a hidden agenda, a physics safe for Catholic belief, and Duhem's was thus 'the philosophy of science of a believer', and that despite the fact that at no point did any of Duhem's arguments rest on Catholic belief. In the same way, Duhem's treatment of the Galileo case seemed to many to have the hidden agenda of restoring the credibility of the Catholic ecclesiastical authorities. From one point of view, talk of Duhem's hidden agenda is most unjust: he never concealed his Catholicism or his political opinions at any time of his life, so much so that from many points of view he was a political innocent, and a letter to Blondel cited below[15] acknowledges as much. But the issue cannot be avoided: it has been, implicitly or explicitly, persistently raised by his critics, and so must be on the agenda of the writer on Duhem.

In contrast the question of Duhem's overt message is more straightforward, though pressing nonetheless: the existence or otherwise of an underlying coherence behind Duhem's varied writings and concerns. On the face of it there is no immediate link between the theory of heat and fourteenth-century philosophical nominalism, between electromagnetic theory and the logic of experiment, between hydrodynamics and the ethical value of obedience. Duhem, however, concerned himself with all these things. Duhem himself used to talk of the Ariadne's thread[16] that, guiding the reader through the Daedalian maze of the world or of his own concerns, would bring clarity and order out of the seeming chaos. Much current writing on Duhem seems to suffer from the lack of that Ariadne's thread.

It should be clear that the answers to such questions are not likely to be found in detailed studies of particular aspects of Duhem's life, work and career, valuable and necessary as these are. There is now a substantial body of work of this kind, and I shall have occasion to refer to it in the course of this essay. Examples are the investigations of H. W. Paul[17] on the lay and ecclesiastical institutional context of Duhem's career, and those of D. G. Miller[18] on Duhem's physics, on all of which S. L. Jaki's biography[19] supplies much valuable information, supplementing the older works of Jordan and Duhem's daughter Hélène.[20] All of this work is in need of extension, particularly in the latter area where there is a pressing need for a detailed examination of Duhem's physics to give him his due place in the history of late nineteenth-century physics. But necessary as these enterprises are, they do not supply the Ariadne's thread that is needed. It seems best to pose the problem from the standpoint of R. G. Collingwood:[21] What questions are being asked? What answers are being given? The questions and answers are those of Duhem, of those who have commented on his work in the past, and of myself. With the questions and answers go assumptions about what sort of questions can fruitfully be asked and what sort of answers can legitimately be given. What we have to consider here is this: what sort of questions, asked by his historian, are going to yield a plausible account of Duhem's questions, and thus enable the historian to begin to make sense of the pattern behind Duhem's work and deal with the nagging worries about his hidden agenda.

Duhem offered his own candidate for the Ariadne's thread I am postulating in 1913, when presenting his credentials on his election to non-resident membership of the Académie des Sciences.[22] Presenting himself as the physicist who had been got by his teacher Jules Moutier to love the theories of physics, Duhem declared that the aim of physics was to give a systematic, ordered, abstract, mathematical, detailed description of the phenomena, making, as remarked above, no appeal to hidden molecular motions. The bases of this approach were tested in logical investigations into the aim and structure of physical theory, and in historical investigations into the past of physical theories. Neither the logical nor the historical investigations had given him any reason to doubt the rightness of his preferred approach to physics, which in any case was vindicated by its fruitfulness in practice: neither had it given him any reason to prefer mechanical or molecular approaches to the subject. In 1941, in a similar, though contrary, vein, Armand Lowinger[23] saw Duhem's philosophy, not as a test of Duhem's approach to physics, but as apologetics for it. Taking a broader perspective, Roberto Maiocchi[24] has related Duhem's physics to *both* nineteenth-century rational mechanics *and* French positivism.

Maiocchi's highly plausible claim is that the Duhem who had been induced by Jules Moutier to *love* the theories of physics developed his own brand of realism in the face of an anti-theoretical philosophical environment engendered by positivism. For Maiocchi, the vindication of the rights of theory over against experiment is the essential theme of Duhem's work, whether in physics or in philosophy. In consequence, contrary to the usual account of him, Maiocchi's Duhem becomes a rebel *against* the instrumentalism of his time, rather than the supreme exponent of it. Within this approach to the interpretation of Duhem, Maiocchi has no difficulty in finding a place for much of Duhem's historical work. Brenner likewise starts with Duhem the physicist.[25]

There is a lot to be said for approaches that start from Duhem's commitments as a physicist, particularly when, as in the case of Maiocchi, they are refined by taking into account the philosophical environment within which these commitments were developed and maintained. It is enough to recall Duhem's refusal to let his name go forward as a candidate for the Collège de France history of science chair in 1904 (the chair to which the crystallographer Grégoire Wyrouboff was eventually appointed over the head of the front runner Paul Tannery) on the grounds that he was a physicist and would go to Paris as a physicist if he went at all.[26] Moreover, it is not difficult to relate all of Duhem's work in the philosophy of science, and much of his work in the history of science, to his concerns as a physicist. But, still, despite the authority of Duhem himself, it will not do. At the very least it will not do for any treatment of Duhem that proposes to take into account how his contemporaries perceived him. It seems to have little to contribute to an understanding of Duhem's persistent habit of quoting or alluding to Pascal; nothing towards an understanding of the weight of evidence pointing to the rôle of religious and political concerns in his life, and nothing towards an understanding of the major shifts in his interests and concerns in the course of his career, shifts which he perhaps never had the leisure and peace of mind to appreciate at their true importance.

It is even possible to doubt whether testing or defending his approach to physics was the only concern behind the writing of the *Théorie Physique*. The doubt is difficult or impossible to resolve in this case, but quite unavoidable when it comes to the later historical work. As will be explained in Chapter VI, though Duhem is now most widely known for his work in the history of mediaeval science, none of his work before April 1904 shows any trace or knowledge of such a thing: the work from that date stemmed from a surprise discovery in the previous winter, and is on a scale far beyond what can easily be explained by his interests in physics. The Académie document mentioned above was written in 1913: in the same year the first volume of his *Système du Monde* appeared, and by then the fifth volume, with its closely argued account of mediaeval philosophical and theological Aristotelianism, was at least in draft if not already written. It is hard to see what Duhem's concerns in physics have to do with the chapters in that volume, of around a hundred pages each, on Albertus Magnus, Thomas Aquinas, and Siger of Brabant. Moreover, even if such material could be successfully related to Duhem's concerns in physics alone, the hidden agenda problem referred to above would be left untouched: the Ariadne's thread has to be sought elsewhere and, it seems, there is nowhere left to look but in the religio-political area. Duhem's Catholicism lies at the heart of the hidden agenda problem: the issue has to be met head on.

4. The Catholic Apologist

As will appear below, Duhem was a citizen of a deeply divided nation, with the Roman Catholic religion he professed at the heart of its divisions. By his study and career choices Duhem put himself into a peculiarly sensitive position in their midst: a teacher of science when those opposed to Catholicism looked to science, and the account positivist philosophy presented of it as their chief intellectual weapon: as a teacher a servant of the state educational system that was a key element in the anti-Catholic programme of displacing religious authority from the national life: as a servant of the state educational system a civil servant owing allegiance to the government of the day, the government that pursued the anti-religious policies he as a Catholic opposed, and whose legitimacy he probably doubted in any case. There was no chance that he could have survived either as a student at the École Normale, that élite nursery of university teachers,[27] or subsequently as a teacher of physics, without an apologetic, without a considered strategy for handling the conflicts he was exposed to.

For coping with conflicts of this sort there are two basic apologetic strategies available to intellectuals, strategies which may be roughly characterized respectively as rigid authoritarianism and non-authoritarian open dialogue. Neither are problem-free, and mixtures of the two are likely to be met with in practice. The first of these is the strategy of both traditional Catholicism and Protestant fundamentalism, but open alike to believer and infidel. Basically a polarizing strategy, it depends on the availability of a clear unambiguous authority that has laid down what is right and wrong, permitting a clear distinction between the friend who accepts this authority and the foe who does not. With it go arsenals of arguments to vindicate that authority against the assaults of the foe, and in the case of Catholicism in Duhem's time, largely provided by the scholastic revival centred on the teachings of St. Thomas Aquinas. It has long been accompanied by traditionalist conservatism, often of a reactionary sort. This strategy has the advantage of clarity: it avoids the fudges so characteristic of democratic politics. Clearly, it was available to Duhem and potentially attractive besides to one of his combative disposition. There is much evidence to suggest that this was the strategy he actually adopted, and it has become usual to interpret his career in such terms.[28]

However, the evidence is not all one way,[29] and in its extreme form this seems an almost impossible strategy for one in his position, while there was much in his situation to steer him towards a more open, less authoritarian strategy, to lead him into serious dialogue with those of his contemporaries with whom he was in profound disagreement. As a university professor, Duhem had of necessity to engage with the thinking of people who would most often not share his faith: as a student at the École Normale, a tiny élite community rarely more than 200 strong,[30] this necessity would have been particularly pressing: and as the author of books intended perforce for a predominately infidel readership, Duhem had *perforce* to allow for the differing commitments of that readership. In yielding to these imperatives, Duhem was of course taking risks. The road of dialogue with the foe, the precondition of any attempt to persuade them to take your beliefs seriously or even adopt them, is dangerous for committed believers: they may end up persuaded instead of persuading, or, even worse, their fellow believers may think this has happened and excommunicate them. But high as the risks are, I hope to show that Duhem was increasingly involved in them, and suffered in consequence. The evidence for the involvement and its results can be seen in both his philosophical and historical works and their subsequent fate. If we are to understand these, we have to take seriously the possibility that to cope with Duhem's Catholicism, it is not enough merely to refer to the fact, nor even to assume that neo-Scholasticism and political conservatism adequately cover the options: we have to look at the evidence of how Duhem interpreted his Catholicism, and be alive to the nuances of an extraordinarily complex period.

In the following chapters I attempt to ground Duhem's apologetic enterprise in a schematic, account of the politico-religious divisions of republican France and their intellectual background: the positivistic scientism of the Third Republic; the authoritarian Catholicism of the Scholastic revival; the reactionary royalism of Charles Maurras's Action Française; the late nineteenth-century Pascal revival. Various kinds of evidence point to Duhem's final rejection of Maurrasianism while increasingly he drew his philosophical inspiration from Pascal rather than Aquinas. On this basis, I offer an interpretation of Duhem's *Théorie Physique,* with a glance at some of the modern debates it has occasioned, before proceeding, after the example of Imre Lakatos,[31] to consider the consequences of Duhem's philosophy of science for his historical writing. An analysis of the surprise that was Duhem's discovery of mediaeval science in the autumn of 1903 will lead to a discussion of the character and implications of historical work that took Duhem further and further from his starting point in physics, and into material whose prime interest was religious and philosophical, and proved, for Catholic and infidel alike, too hot to handle.

There should be no need by now to spell it out: I find Duhem a major figure indeed, and believe that many of the reservations about his greatness are based on a deficient appreciation of his intentions: I find the *Théorie Physique*, though flawed, a flawed masterpiece: I find the path from philosophy to history by no means direct and smooth, and the relation between Duhem's work and the policies of the Catholic ecclesiastical authorities by no means the obvious one: finally I find Duhem's historical work, astonishing in its day, still relevant to modern debates about the rôle of theology in the origins of modern science and philosophy. Read with proper care, he usually turns out to have seen farther than his critics.

II

A Catholic in a Hostile World

1. Non-Philosophical Problems for French Catholics

Sir Karl Popper has remarked[1] on the importance of Non-Philosophical problems in philosophy: for him indeed "Genuine problems are always rooted in urgent problems outside philosophy, and they die if these roots decay." Though Popper is here chiefly concerned to emphasize the specifically *scientific* roots of philosophical problems, he also allows for other types of non-philosophical problem such as those of mathematics, morals, and politics. The point is certainly relevant to Pierre Duhem, who was never short of urgent non-philosophical problems. Not a philosopher, but thanks to the traditional disciplines of French schooling well able to handle philosophy, he was born into and lived his life in an intellectual and political world in which his beliefs and commitments were under perpetual challenge. In an Anglo-Saxon environment one's beliefs and commitments may be a private matter, and what one does as an academic usually of no political interest: not so in France. In that environment defending one's beliefs, devising an *apologetic*, was a matter of survival. A very rough sketch of that environment should help readers understand Duhem's position.

The France[2] Duhem was born into called itself an Empire—the Second—whose emperor Louis Napoleon, styled Napoleon III, like his uncle Napoleon Bonaparte, had come to power by coup d'état. The France Duhem grew up in and made his career in was a Republic—the Third—the regime that, proclaimed in 1871 on the collapse of the Second Empire after the disastrous defeats of the Franco-Prussian War, was to last, to the grief of many of its citizens, up to the fall of France in 1940. For France could easily have been a monarchy instead, ruled by one Henri V, a Bourbon of the family of the Louis XVI guillotined in the Terror: those who set up the Third republic meant it to be provisional—provisional, that is, until full agreement could be reached on a Bourbon restoration. But agreement was never reached, and 'c'est le provisoire qui dure', the provisional Third Republic endured through crisis and scandal for 70 years.

Republican government for France was not inevitable, no more inevitable than lengthy terms of office for the short-lived governments of the Fourth Republic in the years following the Second World War. The French nation was deeply divided, and its divisions were reflected in the succession of regimes that sought to govern it: the first century after the end of the *ancien régime* was marked by three republics, two monarchies, and two empires. Since then, outside observers may have become accustomed to think of French politics as inevitably republican, but even now this is an illusion: convinced royalists are still to be found. How much more would it have been an illusion in the latter part of the nineteenth century when the Third Republic struggled from crisis to crisis and every scandal exposed yet again the depth of the chasms separating Frenchmen from each other! In some countries politics concern the rise and fall and actions in and out of power of particular individuals and political parties: in France the regime itself was at stake. In such circumstances, politics are liable to involve loyalties and values at a basic level, and to have ramifications far beyond the merely political world, and so it was in France. It thus becomes crucial to determine where Duhem stood in this bitter dispute.

He was a Catholic: it was to be expected that he would be hostile to the Republic, for religion was at the centre of the dispute. Though never entirely trouble-free, the intimate alliance between throne and altar had been an essential element in the *ancien régime*. Carried through by people who were often, like the deist Robespierre, committed freethinkers, the Revolution left Catholics hostile to Republican politics, to the politics, as they saw them (and experience often seemed to bear them out), of freethinkers, Freemasons, and Jews. How could it have been otherwise? These, with the Protestants, were the groups who most benefitted from post-Revolution religious freedom.

Duhem's family was described by his daughter as legitimist,[3] that is, it supported the main 'legitimate' branch of the Bourbon dynasty, that of Louis XVI, the branch that had been restored in 1814–15 only to be deposed again in the so-called 'July days' of 1830. It was thus opposed to the rival Orléans branch, which, after siding with the Revolution, had ascended the throne in 1830, only to be driven off again in 1848. Not Bonapartist either, the Duhems did not hanker after a Third Empire ruled by one of the relatives and descendants of Napoleon. All three options appealed to Catholics of different persuasions.

Those who set up the Third Republic had been monarchists who believed they had an electoral mandate to effect a legitimist restoration, of Henri, Comte de Chambord. But they were frustrated by a complicated dispute over the flag of the restored monarchy. The provisional Republic lasted and soon came under the control of convinced republicans pursuing anti-religious policies: persecution of religious orders and building up a national *lay* educational system free of religious control or influence.[4] This turn of events is easy to understand: when the regime itself is at stake, disputes become loyalty tests, not matters on which loyal citizens can agree to differ. While education ministers like Jules Ferry sought to ensure a loyal republican electorate by ensuring the 'right' of every French child to be taught only what had been 'proved', Catholics were ready to use every crisis to destablize the Republic, even at the cost of some highly dubious alliances.

Particularly instructive in this regard is the Boulanger affair of the 1880s, which helped to set the mould for Catholic reactionary politics over the next half-century.[5] Centred on a popular general of that name, his term of office as War Minister, and the events that followed, it is the story of a challenge to the republican parliamentary constitution that originated in the party which then embodied extreme republicanism, the Radicals. That challenge was, however, only maintained thereafter by subventions from the Catholic Royalist aristocracy on the basis of secret 'understandings' that Boulanger would pave the way for a Bourbon restoration. This particular alliance was surrounded by equivocation and embarrassment on both sides: that with Charles Maurras's Action Française in the early years of the following century was another matter altogether.[6]

As will be seen, Maurras was ideologically largely at one with the positivists who supported the Republic. An atheist himself, he conceived a use for the Catholic church and its monarchism in his vision of a revived positivist France, and so could openly appeal for and receive Catholic support, even when his movement had been condemned by the Vatican (in 1926), and do so with a degree of success altogether beyond Boulanger's reach. Maurras owed his success in this regard, though, to the Dreyfus affair, the product as well as the cause of one of those ideological switches that make intellectual history such a fascinating and challenging business.

As is well known, Dreyfus was an army colonel, tried and exiled to Cayenne on fabricated charges of espionage, who became the focus of a long political campaign, aimed at his release, that split France from top to bottom, with republicans mostly on one side and Catholics on the other. That a Catholic like Duhem should want in these circumstances to defend the honour of the army—which, it seems, he did—was by no means self-evident.[7] A generation earlier, loyalty to the army had been something for republicans, and one of Boulanger's initial triumphs had been the exclusion from its officer corps of the Duc d'Aumale, despite his service record, on the mere ground of his royal blood. But in the 1890s things had moved on: the army now stood for order and traditional authority: Catholics took it for granted that they were committed to these things: in their eyes those who undermined the authority of the army by impugning its integrity were the assorted freethinkers, Freemasons, and Jews who, as they saw it, had brought such harm to France, the monarchy, and the Church since the Revolution. For such people Dreyfus's innocence was out of the question. As in all political cases, the other side's evidence was always suspect, for the prime issue was that Frenchmen did not trust each other to act honourably: the only evidence to be trusted was what came from their political friends. The affair ground on, Dreyfus was rehabilitated, and Catholics soon paid for their stance against him when new left-wing governments pursued the traditional Republican anti-religious policies with renewed vigour, dispersing religious orders and finally separating Church and State. A decade earlier, the Vatican had tried to avoid this kind of conflict by encouraging Catholics to make peace with, to 'rally' to the Republic, but all these efforts were now brought to naught, and French Catholics were more disaffected than ever. It is small wonder that many of them rallied to the antisemitic atheist Maurras instead. In such a polarized political atmosphere a natural assumption is that a Catholic and patriotic Frenchman like Duhem would be among these many: his daughter Hélène said so in 1936, and the evidence of his files indicates that the superiors reporting on him thought so too, and the fact of 'Israelite friends' like Jacques Hadamard does not necessarily refute this analysis. But at least one early biographer thought there was more to be said. The daughter's express protest[8] is not decisive: there were Catholics, even patriotic ones, who did not swim with the Maurrasian tide. The biographer who so annoyed Hélène, Édouard Jordan, did so by suggesting that he had some sympathies for Christian Democracy, adding that many of his friends were involved in it (a fact easily confirmed from a listing of Duhem's correspondents). The issue cannot now be resolved, but something about Christian Democracy and its implications may serve to prepare the ground for the following chapters. Inevitably I paint with a very broad brush.

2. Christian Democracy and an Opening to the Left[9]

Rightly or wrongly, it is commonly observed that the rise of modern Western parliamentary democracies was intimately connected with the rise of Western industrial economies, that the people who fought for representative democracy and those who created the factory system were one and the same. Equally common is the observation that the rise of the factory system was associated with much misery and oppression for the people employed by it, who lived in the associated urban slums. The truth of such observations does not concern me here: what does concern me is their triteness and wide currency, voiced equally by Catholic critics of parliamentary democracy in, for example, France and Italy, and by Karl Marx and his followers. The motive in the first case was no doubt similar to that behind the much discussed 'liberation theology' of modern Latin America: the Church would take its cue from much in both Old and New Testaments and 'opt for the poor', make common cause with the labouring masses against their oppressors who, in nineteenth-century France and Italy, just happened to be the same as those who were felt to be oppressing the Church. This attempt to turn the tables from below on anti-clerical lay regimes involved a major break with tradition, the nearly consistent tradition of reaching the masses 'from the top down', via their rulers, the 'secular arm' traditionally required, for example, to burn at the stake heretics condemned by the inquisition. This deeply-rooted tradition had proved itself in the Reformation era, when the assiduously cultivated support of princely houses was often the key to turning back the tide of popular Protestantism and giving the Jesuit and Capuchin missionary orders room to work unhindered, a policy that came unstuck only where, as in England, cultivating the support of the Tudors was incompatible with cultivating that of the Habsburgs, the house to which Henry VIII's first wife, Catherine of Aragon, belonged.

The 'top down' tradition was not, though, quite universal. A 'bottom up' approach was applied, with decisive consequences for later Irish and Scottish history, by the Jesuits in Ireland and parts of the Scottish Highlands and Islands. Appealing to the masses over the heads of their rulers was not, then, quite without precedent, but it was a decidedly risky course, risky not just because of the chance that Italian Catholics, for example, would be forced into the unlikely claim that rule by the Bourbons in Southern Italy was an improvement on rule by their bourgeois successors, but because it was always liable to clash with the alternative 'top down' tradition, which continued to operate, and was always liable to generate inappropriate authoritarian stances and impossible dilemmas for many sincere Catholics.

The problem was to determine the prime objective. In the Italian case examined by Poulat, for the young Italian priest Umberto Benigni it was going to be the assertion of Papal authority: for him the new lay Italian kingdom had both usurped Papal authority and created much misery with its Manchester-inspired social policies. Led by the Pope, the Church had sound social doctrines which the people ought to acknowledge if they wanted salvation from this lay tyranny. That approach *may* have cut *some* ice in Italy, but in France it cut none at all: the French masses were as estranged from the Roman Catholic Church as the English lower orders from the Church of England. With many of them soon to opt for Communism, they were unlikely, in any numbers, to acknowledge the authority of the Church.

In these circumstances, it is no surprise that many concerned

Catholics abandoned the authoritarian approach in favour of dialogue: the Church might in principle be right, but it was pointless to insist on the fact if you wanted to work alongside the poor and oppressed—and alongside non-Catholics doing the same. Down that road went Marc Sangnier's Sillon, an ambitious and successful youth movement condemned by the Roman authorities in 1910,[10] and the Semaines Sociales of Henry Lorin. Both were suspect in the eyes of many Catholics, particularly those influenced by Action Française: they seemed to compromise the authority of the Church and the integrity of its teaching. Hélène implies that her father was hostile to such movements and is able to cite his apparent approval of Maurras's polemical *Dilemme de Marc Sangnier*.[11] She claims further that Duhem was at least sympathetic to Action Française and, notwithstanding his 'Israelite friends', antisemitic in his attitudes. But the fact that one of the latter was the mathematician Jacques Hadamard, brother-in-law of Alfred Dreyfus and principal organizer of the campaign for the revision of the case, and that the friendship was maintained throughout the affair (though the letters do not discuss politics), should give us pause. It also seems[12] that Hélène was herself a supporter of Action Française when she wrote the biography of her father, and so could have had an interest in slanting her story that way. Such indications cast doubt on the prevalent interpretation of Duhem as a conservative reactionary, but do not help us much with determining Duhem's actual position: the correspondence with Maurice Blondel allows us to be more definite about his attitude to the Semaines Sociales and to the controversy surrounding them.

The Semaines Sociales consisted basically of annual jamborees on social questions. When in 1909 Bordeaux was the venue the event brought Duhem a visit from his old student friend, the now influential Catholic philosopher Maurice Blondel of Aix-en-Provence. Soon afterwards a series of articles signed 'Testis'[13] appeared in the venerable and influential Catholic periodical *Annales de Philosophie Chrétienne*, edited by the Oratorian priest Lucien Laberthonnière, which had carried Duhem's famous pieces 'Physique de Croyant' and *To Save the Phenomena*. Along with Blondel and many of their friends, Duhem was listed on its cover as a "principal collaborator". As always with anonymous publishing, the guessing began at once: who wrote it? Laberthonnière was an obvious suspect. Equally, in view of his well-known connection with the *Annales*, Blondel was another. In a letter expressing pleasure at the visit, Duhem sought enlightenment from Blondel. Their letters are a delightfully teasing, cryptic stage of a fascinating correspondence between two intimate friends who understood each other at a deep level.

In these articles, Blondel, for he it was, addressed the question of whether an alliance with Action Française was theologically even thinkable. A variety of issues were at stake: an alliance with Action Française was ultimately an option for the powerful, not the poor, an alliance between the Church that in the Magnificat praised the God who pulled down the mighty from their thrones, and the movement that congratulated that same Church for so wrapping up that same text liturgically that nobody noticed its political content. But the core of Blondel's complaint about the enemies of the Semaines Sociales, who were also his enemies, was the reduction of Christian obedience to mere obedience to human orders: in Blondel's eyes that destroyed human freedom and spontaneity before the Creator and Saviour. There was no question where Action Française stood here: not for the *freedom* of conscience but for its *direction*.

Though elusive, cryptic and playful, the correspondence is decisive: Duhem sided with Blondel, not with Maurras and his Catholic supporters, and was if anything cheering Blondel on. But in any case, in 1916 Duhem spelled out his own position on the subject clearly enough for those with eyes to see, in a piece of widely-circulated War Literature, 'Science allemande et vertus allemandes'.[14] Here, unsurprisingly, the German virtues are said to be implicit and absolute obedience, whether to men or to scientific deductions—Duhem links the two together in a way that seems, for reasons to be discussed in the next chapter, to be characteristic of much Catholic thought. In Duhem's eyes both were only virtues up to a point. Absolute obedience to scientific deductions was not a virtue if the results were an affront to common sense or otherwise absurd. Scientific progress needed creative spontaneous intelligence. Absolute obedience to orders was obviously not a virtue if the orders were cruel and inhumane and in general it was only a virtue in a monk obeying the Rule of his Order as laid down by a saint.

Did those demanding implicit obedience in Duhem's day meet a like condition? The sanctity of those who governed the Church in Duhem's day has since been affirmed by the canonization of Pope Pius X. Duhem, though, may have thought otherwise. As a result of Blondel's campaign the *Annales de Philosophie Chrétienne* was denounced to Rome and put on the *Index* in 1913. In a postscript to a letter to Blondel during the resulting crisis Duhem wrote: "Oremus pro Pontefice Nostro Pio. Dominus custodiet eum et conservet eum a malis circumstantibus." (Let us pray for Pius our Pontiff. May the Lord guard him and protect him from the evil people around him.)

3. The Positivist Challenge and Duhem's Response

In the opinion of Jules Ferry, it was the right of every French child to be taught only what had been proved. There might well be several opinions about what had been proved, but there can be little doubt about what, generally speaking, Ferry had in mind, and it did not include the Catholic faith. As has been mentioned above, Ferry had set out, with his predecessors and successors, to build on the revolutionary inheritance a national *lay* educational system free of clerical control or influence, and as a university professor Duhem was to serve that system. By this programme, the republicans could hardly have signalled more clearly that in their struggle against ecclesiastical power and authority their strategy looked to intellectual weapons, weapons largely derived from the scientific and quas-scientific themes of the time, a strategy largely shaped by the Catholic tradition, to be considered below, of intellectualist apologetics.

It is not difficult to imagine the kinds of argument the republicans were hoping to deploy: many are still in use today. At the factual level, science could be used to ridicule Biblical and more recent miracle reports, and transubstantiation, while evolutionary theories could be used to undermine the rationalist pretensions of natural theology. It could be suggested that theology and the Bible offered credible explanations only in areas scientific progress had not yet reached, and reach them it soon would. A materialist theory of life could render implausible or impossible belief in an afterlife: and physical determinism could be used to undermine Christian ethics. Arguments of another kind related to ethics and morality: to those who felt that Christianity was required to provide a foundation for morality, the apostles of the religion of science, like the physical chemist and politician Marcellin Berthelot,[15] could suggest that morality too could be founded on the theories and practice of science.

More serious, though, were those challenges that appealed to philosophy and history: science was more rational in its beliefs and practices, and in the past the Church had obstructed the progress of science. Both, of course, were already implicit in the particular 'factual' and 'ethical' challenges to the Catholic faith mentioned above, but in their explicit forms they were more dangerous and difficult to deal with. One way or another, both purported to highlight the incompatibility of Catholicism and science. A good summary statement of both is provided by

Duhem in a letter of 1911 to J. Bulliot, Professor of Scholastic Philosophy at the Paris Institut Catholique, and reproduced by Hélène in her biography,[16] a letter that will play an important part in the argument of this essay.

Duhem's letter addresses these two principal issues: the alleged continual opposition of the Church to scientific advance in the past, and the alleged superior rigour of scientific proof compared with religious belief. For Duhem it was all *lies*: Catholic opposition to the progress of science in the past was a lie, and equally a lie was the superior rigour of scientific proof. The main interest for this essay lies in the grounds on which he chose to defend these two claims. By 1911, the main features of Duhem's response to the first, *historical*, challenge were in place, but only recently so: their development will be considered in the latter part of this essay. It is hard, however, to know what he would have said about it in the 1890s. At that stage the main elements of his response to the *philosophical* challenge, as he perceived it, to his Catholic faith were being assembled and deployed with increasing effect—though there were to be important changes of emphasis over the following decades.

4. The *Non Sequitur* Apologetic

Famously, Duhem's response to this second challenge, the supposed superior rigour of scientific proof, was '*non sequitur*', it does not follow: to him, it was a matter of logic that in physical theory, and indeed in other branches of science, experimental proof was incapable of yielding the anti-religious results claimed. In one sense, Duhem's claim was hardly news: questions of experimental method and proof had been debated since the seventeenth century at least, and with particular intensity throughout the nineteenth. It was widely agreed that experimental proof could yield no certain information about the inner nature of things, about how things truly were in themselves. The conclusion Duhem drew, however, was more newsworthy and debatable.[17] Following Auguste Comte, the tendency had been to say that science should stick to what was called *positive knowledge*, the bare laws of the phenomena and the co-ordination of these into coherent wholes, avoiding questions about their modes of production or causes. But Duhem proclaimed even this goal utopian: there was no certainty to be had even there, for the mere statement of the laws of the phenomena depended on hypotheses, that is on *theories*, logically speaking purely arbitrary, that experiment in the strictly logical sense was powerless to prove or disprove.

It is important to see what is at issue here. Comte and his *positivist* followers had proposed to get round the impotence of experiment to get at the truth behind the phenomena by, along with many other philosophers before and since, abandoning the search for truth altogether. To use a convenient if overworked example, if, no matter how many white swans you observe, you cannot prove thereby that all swans are white, then, it may be argued, the notion of truth is illusory: we should content ourselves with the phenomena, with what appears, and in our science claim no more than our experiments, in strict rigour, require. If the positivists were asked why anyone should want to pursue such a patently limited form of 'knowledge', the answer would have been that it was useful in classifying our experiences and in arranging for the necessities of life and effecting improvements, that it was a useful *instrument* for meeting human needs. By showing the necessary involvement of theory in the mere statement of experimental laws, theory that no amount of rigour could ever eliminate, Duhem turned the flank of this *instrumentalist* (the standard term) methodology, and thereby paved the way for the rehabilitation of the search for truth in the twentieth century, notably in the work of Sir Karl Popper. Duhem's possible *realism,* belief in truth, then, is one of the principal themes of this essay.

In view of my remarks above about the rôle of non-philosophical problems in philosophy, it is no surprise that the positivist instrumentalist programme had wider applications. The philosophy of science of Comte and his various followers served, and was intended to serve, as a model for a science of society: just as the science of non-human things consisted of purely descriptive laws of phenomena to be judged by their usefulness, not their truth, the science of human affairs too was to consist of the laws of social phenomena only, to be judged by their usefulness, not their truth, for, as has been remarked above, truth for the positivists was a meaningless concept, and into the same dustbin questions of right and wrong were as swiftly discarded as questions of truth. 'Usefulness to whom?', we might well ask: usefulness to the positivist élite proposing to use its knowledge of the laws of society to remould that society and improve it according to their lights, whatever the opinions of individual citizens on the matter.

The effects of all this can be seen in the treatment of religion by the positivists: "The various modes of worship which prevailed in the Roman world, were all considered by the people as equally true: by the philosopher as equally false: and by the magistrate as equally useful."[18] The would-be magistrates of the positivist persuasion, who particularly prized social stability, convinced themselves that religion was needed in the interests of that stability. To meet that need, August Comte invented his Religion of Humanity, liberally supplied with rites adapted from those of the Roman Catholic Church. Charles Maurras and his friends in Action Française had, as has been seen, the alternative idea of making use of the Church instead: the normal processes of political coercion and manipulation could ensure that it served *their* purposes. Needless to say, no justification was offered for this 'value-free' socio-political programme: a large empirical induction enabled Comte to announce that human history was encompassed by three stages, religious, metaphysical, and positive, the last of which had now arrived, so that questions of truth could be easily dismissed by consignment to the metaphysical stage now past. Those who persisted with such questions could be brushed aside by the Comtians as hangovers from the past, and their arguments ignored. Not everyone went as far as Comte and his followers, but positivism and its derivatives combined to give a massive push towards a scientific climate in which serious religious views could only with difficulty get a hearing. To a large extent they set the agenda for the intellectual life of the time.

5. A Catholic Objection to Duhem's Views

Roberto Maiocchi has considered[19] in some detail the consequences of this scientistic and instrumentalist climate for science itself in general and Duhem in particular: plausibly, he argues that the pervasive instrumentalism led to anti-theoretical attitudes in which scientists attempted, in complete rigour, to confine themselves to what could be directly *deduced* from experiment, just the kind of enterprise whose failure Duhem recalls in the opening pages of 'Physique de Croyant'. Much of Duhem's early writing is easily seen as a considered response to such attitudes, one in which a demonstration of their insufficiency as a basis for science is used to reassert the rights of theory. Maiocchi's work is a major contribution towards an understanding of Duhem. Curiously, though, it does not consider in any detail Duhem's first published work on the philosophy of science, his 'Quelques Réflexions au Sujet des Théories de Physique' of 1892, which, on the face of it, is as near as anyone could ever get to instrumentalism. In it, physical theory is supposed to be an instrument for assisting the memory, by providing a classification of the bewildering mass of experimental laws, which, in turn, are purely symbolic, bearing to the facts they represent the relation of signs to things signified. Duhem followed this up by applying it to atomistic symbolism in chemistry,[20] showing, to his own satisfaction at any rate, that all the practical and heuristic advantages of atomic notation could be had without any commitment to the existence of atoms. Duhem's decisive turn away from instrumentalism came the following year when he set about dealing with the criticisms of the Catholic civil engineer Eugène Vicaire.[21] At this point he introduced the series of qualifications to that instrumentalist theory that have perplexed Duhem's readers ever since.

The specifically Catholic context of Vicaire's criticisms will be discussed in the next chapter, but it will be enough for now to note two aspects, which independently of any Catholic context, evoked significant responses from Duhem, both at the time and later. The first of these homed in on Duhem's stated view, reminiscent of that of the Austrian positivist Ernst Mach, of the purpose of physical theory: assistance to the memory. Vicaire's comment seems devastating: if assistance to the memory is the only object, then mathematically formulated theory is not obviously the best way of achieving it—a collection of menmonics, subject to no demand for coherent ordering, would seem more suitable. The second was more difficult and far-reaching: Duhem gave no guidance on the selection of hypotheses, and indeed in one sense he never did give such guidance. His account is at every point inductive: we start with the phenomena, and consistently seek hypotheses that are as little as possible in advance of the evidence. Duhem even considers the supremely positivist 'méthode idéale et parfaite', ideal and perfect method, in which nothing, but nothing, is added to the experimental data. Though Duhem recognized that this was actually, and indeed absolutely, impossible, he did not make that concession in strong enough terms to prevent Vicaire pillorying him for it unmercifully.

Duhem's views were constructed from within the circle of assumptions and problems of nineteenth-century positivism. Vicaire would perhaps have liked to force Duhem out of that circle, but in that he was unsuccessful. He would have liked an account of physical theory that in a simple-minded way gave it access to the truth behind the appearances and allowed that fact to influence theory construction, but Duhem did not oblige. Duhem was to deal with both types of criticism in ways that stayed firmly within the positivist scheme of things. At the same time he showed the latter's limitations when taken really seriously, *and* removed Vicaire's hoped-for simplicities forever beyond reach.

Duhem's response to the first of Vicaire's criticisms, the superiority of mnemonics if memory assistance was the only aim, was 'L'École Anglaise et les Théories de Physique', the article that was in due course to form the nucleus of one of the most famous chapters of his *Théorie Physique*. In an evocative presentation spiced with theories of national psychology, Duhem acknowledged *both* that overall coherence in physical theory (which would rule out a mere collection of mnemonic devices) was indeed his aim, *and* that no logical analysis of experiment could ever justify that aim: the only justification he could offer was a hope, a hope that the classifications imposed by the theories on the phenomena by their mere existence were ever converging on an ultimate classification, whose final form remained unknown. In an implicit appeal to eighteenth-century debates on biological classification, Duhem called this a *natural* classification, thereby introducing an idea that was to be a permanent feature of his thinking.

This embryo anticipation of Popper's theory of verisimilitude, of approach to the ultimate truth by way of partial truths,[22] is hardly startling. But Duhem's response to Vicaire's second count against him was more radical: a comprehensive critique,[23] that soon became a classic, of the whole notion of a purely experimental physics. This shaped all subsequent discussion of the topic, and took its final form as Part II of the *Théorie Physique*. The details will be discussed in their place, but it is enough for now to note the overall result that theory is necessary for experiment, necessary to set up the experiment, and necessary equally to describe its outcome: theory-free experimental physics is a chimera, an impossibility. That being so, experimental physics has either to be given up, or theory given its legitimate rights. Duhem's devotion to theory over experiment may have been one source of his career difficulties, and the experimental work he encouraged some of his Bordeaux students to undertake seems to have been theory-led. Maiocchi is surely right in his claim that the defence of theory was Duhem's principal aim, and not too far-fetched either in his suggestion that this piece, like the 'École Anglaise', is an attempt to argue for realism on positivist premisses by, in effect, reduction to the absurd. Vicaire's criticisms were the occasion of these developments, but they remain, in their anti-instrumentalist effects, expressions of his fundamental commitments.

But it is when he considers the question of the source of the hypotheses of physics that Duhem's response to the second objection meets that to the first, just as the grounds for seeking a coherent physical theory lay in an aspiration to a natural classification that experimental method could never justify, so the hypotheses did not come from the experimental situation, but from history and, as Duhem was later to make clear, they were not a matter of choice at all, but just grew in the minds of physicists.

With that, then, Duhem had provided his answer to the positivist attack on the Catholic faith: like religious belief *both* the experimental method in physics, *and* its practice by physicists, rested on assumptions incapable of proof: common sense or, perhaps, even metaphysics, was needed to justify the demand for coherent physical theories: and the hypotheses of physical theory, necessary as they were for the conduct of physics of any sort, were and would remain unproven. Thus, despite positivist propaganda, physics was in the same case as theology, and the presumed superiority of physics in demonstrative rigour, taken seriously, turned out to be a will o' the wisp. Duhem's case rested on the insufficiency of what had come to be called 'positive science'. It would have been insufficient to show that positive science could not handle non-scientific matters: the positivists knew that already and were glad. Duhem's arguments purported to show that it was insufficient in itself, and for that reason have come to haunt philosophers of science ever since. But achieving that result against the positivists was one thing, convincing his co-religionists another. As I hope the next chapter will show, this was a much more difficult matter.

III

Defending Catholic Authority

1. Difficulties for a *Non Sequitur* Apologetic

Pierre Duhem had constructed arguments purporting to show both the impotence of scientific arguments in the service of the anti-religious cause, and the uselessness of science as a model for 'scientific' intellectual standards supposedly superior to those upheld by the Catholic faith. His system "sweeps aside the supposed objections of physical science to theistic metaphysics and the Catholic faith".[1] On the face of it, such arguments ought to have been welcome to the Catholic Church of his time: it is almost incredible that the deviser of these arguments should not have been fêted everywhere by the representatives of his Church. But the incredible was the case, for reasons having to do with the distinction between authoritarian and non-authoritarian apologetic strategies aired in my previous chapter. In brief: Duhem's arguments did not positively *conduce* to Catholic belief, and to obedience to Catholic ecclesiastical authority. Worse, they tended to undermine widely-used arguments to that end. Duhem's system also "denies to physical theory any metaphysical or apologetic import."[2] The negative apologetic that cut at the root of anti-religious scientism did the same to attempts to use science for the purposes of positive apologetics, for the purposes of natural theology. Welcome as was Duhem's claim that physics rested on unproven hypotheses, the suggestion that theology was in the same case was much less so. The ecclesiastical authorities were looking for arguments not just to repel positivist marauders, but also to help them win the war and defeat the enemy for good, arguments to induce belief and predispose hearers to accept *their* authority. In this they were doing no more than being true to their Church's long history, but as the point has rarely been appreciated by Duhem's readers, it merits further discussion.

In considerations of the precise apologetic import of Duhem's views, it is well to begin with what is perhaps the classic exposition of them, the source of the quotations in the previous paragraph, his 'Physique de Croyant' (Physics of a Believer) of October-November 1905, published in the *Annales de Philosophie Chrétienne* for those months,[3] and in due course included in the second edition of the *Théorie Physique*. The occasion of this piece is well known: the article in the *Revue de Métaphysique et de Morale* of the previous year by the young philosopher Abel Rey[4] in which, using work that reappeared in his 1907 thesis,[5] he argued that in its overall effect Duhem's was the philosophy of science of a believer. The main factor in this conclusion was Duhem's denial to physics of any metaphysical import, any power to say anything about the world as it really was, just the feature in fact that removed from physics its usefulness for anti-religious purposes. Curiously, Rey's argument had set aside the long tradition of positivist instrumentalism discussed above, the property of a group hardly likely to be accused of religious tendencies! It also betrayed an ignorance of the more popular forms of Catholic apologetic, and a failure to consider what Duhem's views might do to attempts, which were legion, to use science to *advocate* Catholic faith. For, if physics was useless as a weapon against Catholic belief because it could say nothing directly about reality, *by the same token*, it was equally useless as a prop for Catholic belief: it was neutral in the conflict, irrelevant.

One example of the sensitivity of the issue is the storm in a wineglass caused by Ferdinand Brunetière's reference in 1895 to the "bankruptcy of science".[6] From all sides good men and true rallied to the defence of the honour of science so lightly impugned, and a banquet was held the better to sing the praises of science. Those present naturally included republicans like Duhem's old enemy Berthelot, but there also was Maurice d'Hulst, Rector of the Paris Institut Catholique: suggestions that science was bankrupt were as serious a threat to his vision of Catholic apologetic as they were to the anti-Catholicism of his republican opponents. Another example is the storm caused by Maurice Blondel's 'Letter on Apologetics' of 1896.[7] In his thesis *L'Action* of 1893 Blondel had proposed a "method of immanence" in which the apologist was supposed to start from the position of the subject who might or might not believe. He attempted to argue that the existence and life of such a subject (technically referred to as 'action') presupposed a transcendent God. In 1896, from a generally similar point of view, Blondel now argued the radical uselessness of science for apologetic purposes.

As it happens, we have Duhem's opinion of Blondel's piece, in a letter of the following year.[8] It had come as no surprise to him: they had discussed these matters as fellow students at the École Normale, and mentioned them in previous letters. Duhem had also received a complimentary copy of Blondel's thesis.[9] Duhem now considered Blondel's argument consistent and valid, provided its premisses were first accepted, provided it was first supposed that the method of immanence was the right or only possible approach to apologetics, and Duhem was not sure that it was: it seemed to him, even in 1897, that Blondel assigned too restricted a scope to science in general and physics in particular. Duhem did not say here what he had in mind, but it seems that what may have been at issue was the possibility of a natural classification, in which the classification of the physicists progressively approached that of the ontological order. But that was perhaps a minor reservation: two years earlier, at a Catholic international congress in Brussels, Duhem had created a sensation by his broadsides against the attempts of Catholic clerical philosophers to use science for their apologetic purposes.[10] One of his victims, Albert Farges, was convicted of equivocations in the course of his argument. Another, Jean Bulliot, was accused of attempting to make use of science without the long prior intimate acquaintance with it that was the essential precondition for the enterprise, if Catholics were not to be laughed off the stage by better-prepared positivists.

But if in 1896 Duhem still harboured hopes for scientific apologetics, and was resisting Blondel's contrary thesis, in 1905 he was agreeing with Blondel: his account of the method of physics removed from it any apologetic import whatever. It was the philosophy of science of an unbeliever quite as much as that of a believer. I have already mentioned the sensitivity of that position from the Catholic point of view. It needs, however, futher exploration, for the issue thereby broached stood at the centre of the religious crisis, the so-called modernist crisis, of the time.

2. Fideism or Rational Obedience

In the encyclical *Pascendi Dominici Gregis* of 1907,[11] two years after Duhem's 'Physics of a Believer', the official position was made clear in the name of Pope Pius X. Of the dangerous aspects of the heresy it called "modernism" identified by the encyclical two concern me here, what it called the "agnosticism" of the modernists, and the separation of science and faith. The first for example was dangerous because of the damage it did to natural theology:

> human reason is confined entirely within the field of the *phenomena*, that is to say, to things that appear; it has neither the right nor the power to overstep these limits. . . . Given these premisses, everyone will at once perceive what becomes of *Natural Theology*, of the *motives of credibility*.

The second was under suspicion of fideism:

> We have proceeded sufficiently far . . . to have before us enough, and more than enough, to enable us to see what are the relations which Modernists establish between faith and science . . . in the first place it is to be held that the object-matter of the one is quite extraneous to and separate from the object-matter of the other. For faith occupies itself solely with something which science declares for it *unknowable*. Hence each has a separate scope assigned to it: science is entirely concerned with phenomena, into which faith does not at all enter; faith, on the contrary, concerns itself with the divine, which is entirely unknown to science.

Pascendi goes on to suggest that the modernists really meant to subject faith to science but were afraid to say so, and was even to find pantheistic implications in the position. It can be assumed that one target of this passage was Alfred Loisy's attempt to separate the results of the critical analysis of Scripture from the dogmatic claims of the Catholic Church, and that another was the memory of late mediaeval and Renaissance theories of double truth, truth in philosophy separate from truth in faith; but the concern of the first passage to preserve the integrity of natural theology shows the importance of wider considerations. Duhem is not one of those identified as targets of the encyclical. Whether his work was even known to those who drafted it must be a matter of speculation. But quite apart from his explicit disapproval of the enterprise, it is hard to envisage the kind of natural theology that could be accommodated to Duhem's account of the aim and structure of physical theory.

The issue was basic, the subject even of dogmatic definition, by the First Vatican Council of 1870–71, the Council that, in non-Catholic circles at least, is more famous for the definition of Papal infallibility. That Council, using the double negatives usual in such definitions, had declared anathema anyone who should deny that the knowledge of God was accessible to human reason.[12] Though what that meant is not easily determined, most of the bishops present must have meant to say that the knowledge of God was accessible to human demonstration, not of course that knowledge of Him contained in the creeds and dogmatic formulations of the Church, but the knowledge that there is a good God who created all things: the rest belonged to Revelation, not reason.

But whatever the bishops thought they were doing thus making the *demonstrability* of God's existence a matter of *faith*, the general strategy is clear enough: to offer the faithful what are technically known as *motives of credibility*, reasons that would make it rational to accept the Catholic faith and ecclesiastical authority. The faithful could be assured that, God's existence being demonstrable by reason, it was rational to accept the dogmatic formulations of the faith concerning Him offered by the church, and rational also to accept the authority of the Church that propagated this faith, rational to support the Church as it defended its temporal power against the new kings of a reunited Italy, and rational also to defend the Church in its resistance to the Prussian rulers of Germany and republican rulers of France. At the same time, of course, it turned Catholics into a disaffected element in all three states, a disaffected element the authorities had to disarm as the price of their own survival.

So it emerges that by propagating a system of physics that undermines natural theology Duhem has rendered himself suspect of the heresy (for that is the effect of the Council's decree) of *fideism*, the belief that the faith rests on faith and nothing else, and that conclusion was explicitly drawn by F. Mentré, one of those who wrote on Duhem's work after his death:[13] in his eyes, Duhem's views were of no use on religion because of this fideist taint, connected with what he identified as its Pascalian sources, the subject of the next chapter. Furthermore, the counterpart of fideism is philosophical scepticism, the doubt about the reliability of knowledge of any kind, about its ultimate guarantees. Thus in 1893 Eugène Vicaire detected in Duhem's views "the poison of scepticism"[14] and was appalled that such views should appear in a Catholic journal, the *Revue des Questions Scientifiques*, in which Duhem's early articles appeared.

3. The Revival of Scholasticism

But the consequences go even further, and it is here that neo-Scholasticism[15] falls to be considered. There is no point in asserting the demonstrability of God's existence, or of anything else for that matter, unless there is available a philosophical system to do it in, just as to demonstrate God's non-existence a philosophical system, such as that provided by the various brands of positivism, was equally necessary. In 1878 the Encyclical *Aeterni Patris* of Leo XIII[16] hit the nail on the head by citing St. Paul in favour of its view that 'false philosophy' was the chief source of the modern apostasy. The 'true philosophy' offered in its place was a revived Scholastic philosophy, the philosophy associated by the encyclical indiscrimately with Thomas Aquinas and the latter's thirteenth-century Franciscan contemporary Bonaventure, but in practice looking more to Aquinas as interpreted by such sixteenth-century commentators as Cajetan and Suarez.

The groundwork for the encyclical had been laid over the previous half-century. At least since the sixteenth century, there had always been a tendency towards Scholasticism in Catholic philosophy and theology, and corresponding difficulties with 'modern' philosophies. For example, an episode better known than some because of the attention Leibniz paid it, is the persecution of the followers of Descartes in the 1670s and 1680s because of the difficulties their philosophy created for the Eucharistic doctrine of transubstantiation defined by the sixteenth-century Council of Trent,[17] the doctrine that at the words of consecration the *substance* of the bread and wine are transformed into the *substance* of the body and blood of Christ, leaving only the so-called *accidents* of colour, taste, and texture unchanged. Despite the intentions of the Fathers of the Council of Trent, this doctrine can hardly be understood outside the philosophy of substance and accident in which it is stated, still less if matter is defined, as it was by Descartes, as *essentially*, in substance that is, the space it occupies—and nobody suggested that that changed at the words of consecration!

Just as at the end of the nineteenth century Duhem was to turn the philosophy of positivism against the anti-religious conclusions of the positivists, so, in the early nineteenth century a series of attempts was made, all of them resulting in condemnations for their authors, to adapt philosophies of non-Catholic origin, such as those of Kant and Schelling, for the purposes of Catholic apologetic. As interpreted by many, the problem seems to have lain in perceived violations of the balance between faith and reason demanded of Catholic orthodoxy: reason was to offer *motives of credibility*, to prepare the ground for faith by making credible the acceptance of a faith such as that revealed in the Scriptures and the decisions of the Councils of the Church, but not to go further. A group of Rome-based Jesuits, of whom the most prominent were Matteo Liberatore and Joseph Kleutgen, argued that only a scholastic philosophy, looking to that of Thomas Aquinas, would do, and they convinced Gioacchino Pecci, the future Pope Leo XIII, of the merits of their case.

In the Pope's mind the main object of the scholastic revival was theological, but there were at least two others: a revaluation of the thought of the Middle Ages, and of the Church's rôle in it, and indeed a re-affirmation of its value in the face of widespread denigration; and the thought that a revived Scholasticism, judiciously interpreted, might be found of use in questions of modern philosophy and science, despite prevalent expectations to the contrary. In short, by this action the Pope intended to re-affirm, against all the apparent odds, what he saw as the Church's heritage, and from that would no doubt flow improved morale among the troops confronted by a hostile world and, even, greater respect for the Church among those who did not belong to it.

The encyclical thus led to a theological programme to re-examine and restate Aquinas's 'five ways' for proving God's existence, to a historical programme to recover the mediaeval materials on which scholastic thought ultimately rested, and to a sustained attempt to apply scholastic thought to contemporary questions, whether social, ethical, or scientific. These different programmes were not of course independent: the scholasticism available to the Pope had been mediated by the work of commentators at least two centuries distant from Aquinas's own time, so that a genuine revival of his thought had to go back past these to the sources, to find out what Aquinas had actually said; Aquinas's work was also apparently embedded in an obsolete natural philosophy, of generally Aristotelian character, so that if his theology was recoverable, it had in some way to be reconciled with later scientific ideas; and any application to contemporary questions had to depend on answers to the prior question of what the philosophy was that was to be applied.

The later progress of scholarship was in due course to call into question most of the answers initially given, but it is these initial answers that concern Duhem: the work of Aquinas was supposed to consist principally in the reconciliation of Christianity with Aristotle, and in the mid-twentieth century Dom David Knowles, a historian with a low opinion of the scholars of the century following Aquinas, was still presenting Aquinas's supposedly successful achievement of this goal as the crowning intellectual achievement of the Middle Ages.[18] It was this Aristotelianism of Aquinas, and of scholastic philosophy as it was generally understood, that was perceived to be the main stumbling block in the way of a Scholastic revival, and it will be a main problem for the remainder of this essay. To the extent that Duhem was involved in neo-Scholasticism, if he was so involved, that involvement can be expected to show itself in Aristotelian themes in his work, and by a sympathetic treatment of mediaeval Aritotelianism, as well as by associations with journals with generally neo-Scholastic policies.

4. Duhem the Scholastic?

Given the obstacles to scholasticism mentioned in the previous paragraph, it is understandable that different would-be scholastics adopted different strategties to meet them. The Roman seminaries, for example, apparently maintained an integral Thomism with few compromises towards modern science, but this was hardly a serious option for a practising mathematical physicist interested in having his work taken seriously by his contemporaries. Another obvious possibility would have been to reject the whole thing root and branch, either outright, or in the manner of the Italian Agostino Gemmelli, who in 1904-05 proposed a Scholasticism that included modern thought.[19] I believe that something like this was Duhem's final position, but it was, as will be seen, decisively rejected by Pope Pius X in his Encyclical *Pascendi* of 1907, and something will be said below about how Duhem got there: uncomfortable as neo-Scholastics felt about his early papers, it is by no means obvious that outright rejection was his position when he wrote them, and his earlier reactions to Blondel seem to point the other way.

An obvious intermediate position was to try to adapt scholastic natural philosophy to make it conform to the discoveries of modern science, the position that seems to have suited the eirenic temperament of Désiré Mercier, the future Cardinal

Archbishop of Malines, and his group at Louvain. This was the programme behind the Société Scientifique de Bruxelles, which Duhem seems to have joined as a lecturer at Lille, and he seems to have appreciated its attitudes and approaches enough to attempt to recruit Paul Tannery into its membership.[20] Apart from its *Annales*, which published his *Les Théories Électriques de J. Clerk Maxwell*, its principal organ was the *Revue des Questions Scientifiques*, a heavyweight quarterly carrying in depth discussions of the scientific questions of the day, but intended for an educated lay audience. It was not afraid of carrying long multipart articles, and of these Duhem became a major contributor: it seems to have been his preferred place of publication for general philosophical and historical pieces throughout the 1890s. This was the journal that carried his 'Réflexions' of 1892, Vicaire's critique, and Duhem's replies. One of the latter, not considered so far in this essay, seems as good a place as any to begin a consideration of Duhem's possible relation to neo-Scholasticism.

'Physique et Métaphysique' of 1893[21] was the first of Duhem's replies to Vicaire. It addressed the suggestion, made by more than one of Duhem's readers, that his rigorous separation of physics from metaphysics was no more than a cover for denigration of the latter: the metaphysician was free to get on with it in his corner while the Duhemian physicist got on with it in his without interference, the implication being the positivist one that physics was the only real knowledge to be had. Duhem insisted that this was not his intention. On the contrary, metaphysics was for him a genuine form of knowledge, indeed "more excellent" than physics, but separated from physics by having different objects and being governed by different methods. The Scholastic expertise with which he set out his views seems to have impressed not a few of his readers enough to make them wonder whether he had a scholastic mentor, for in the normal course of events this was not the sort of expertise a physicist could be expected to have. Be that as it may, Duhem was claiming to be classifying independent and legitimate sciences, not distinguishing sense from nonsense in the manner of earlier and later positivists.

Repeated in Duhem's later writings, this move has been the main source for the view that Duhem's prime philosophical inspiration was neo-Scholastic. But initially plausible as this interpretation may seem, it becomes less so when Duhem is compared with a genuine neo-Scholastic like Jacques Maritain,[22] who did indeed distinguish his sciences, but only so that thereafter he could unite them, assign each of them its place in the overall system of the sciences, and say which sciences could and could not establish what on the foundations of which others. The basis for Maritain's scheme, as of numberless others of like provenance, is the view that some sciences can be subordinated, or subalternated, to others in the Aristotelian scheme of things.[23] A science is conceived of as a deductive system of syllogisms, deduced from one or more definitions of the essences that are the subject matter of that science, and remaining within its genus or natural kind, and it is supposed that the conclusions of one science can serve as principles for another, as when the sciences of equilibria and music take their principles, as *subaltern* sciences, from the superior sciences of arithmetic and geometry. Famously, this scheme ran into difficulties with the applied mathematical sciences, such as astronomy in ancient times and terrestrial physics in modern: if Aristotle was right, natural philosophy should have been subordinate to 'physics', or, in Duhem's terminology, 'metaphysics', but the mathematical science of nature soon left Aristotelian metaphysics far behind, a point that will be considered further below.

I mention now two aspects of such schemes: they were only achieved at the price of distorting Aristotle, for whom mixed sciences would have meant mixing genera, something his methodological principles forbade; and their rationalistic atmosphere, not to say *hubris*, is remote indeed from a scientific world in which, as in the physics of Duhem, mathematical formulae are devised to meet the problems thrown up by experiment, not those suggested or deduced from *a priori* theory. The reconstruction of Aristotle that would reconcile his views to modern science would have to be pretty radical, radical both at the level of method and of content.

Nevertheless, there seems to have been one aspect of Artistotle's system that Duhem found somewhat promising: its freedom from *a priori* selection of the *primary* qualities by which, in the manner of the mechanical philosophy of the seventeenth century,[24] all *secondary* qualities had to be explained. For him, what qualities were primary and what secondary ought to be a purely pragmatic matter, decided by the progress of theory and experiment as successive theories succeeded in classifying wider and wider collections of data. Where, though, he differed from Aristotle is that his physics was to be a mathematical science; it was to classify qualities, not explain them, and do so by replacing their measured intensities by symbols subject to mathematical manipulation; it was to be a mathematical science whose form no metaphysical system could decide *a priori*: that form too was to emerge from the progress of physics, as successive theoretical classifications of the mathematical intensities of qualities, and the implied classifications of the qualities these represented, hopefully converged on the natural classification that was the goal of physics.

Such was the 'Aristotelianism' that Duhem advertised in a variety of articles in the middle to late 1890s, particularly in his historical works culminating in *Le Mixte et la Combinaison Chimique* and *L'Évolution de la Mécanique*, before repeating it yet again in 'Physique de Croyant' after which it disappears from view. It was a pretty minimal Aristotelianism, but after 1905 even that disappears from view. To my knowledge, Duhem never withdrew such views, but their disappearance from his later writings is indicative of his final decisive rejection of neo-Scholasticism and all it stood for, of those of his earlier attitudes that had made it reasonable for Blondel in 1893 to tease him as a peripatetic. Duhem's ultimate reasons for this shift are not completely clear—some possible answers will be explored later in this chapter—yet there can be no question but that it came at a critical time, when the so-called modernist crisis[25] was at its height, and neo-Scholasticism lay at the heart of that crisis.

I have already referred more than once to the Encyclical *Pascendi Dominici Gregis* of 1907: the prominent rôle neo-Scholasticism played in it can hardly escape the notice of any reader; implicit in the doctrinal part, it becomes explicit in the disciplinary part that follows. We are told that a distaste for the scholastic method is the surest sign of modernism in any writer (What else would be expected of an adherent of a modern philosophy but opposition to scholasticism?), and the text goes on to insist that scholastic philosophy is henceforth to be the basis for Catholic thought:

> We will and strictly ordain that scholastic philosophy be made the basis of the sacred sciences. . . . And let it be clearly understood above all things that when We prescribe scholastic philosophy We understand chiefly that which the Angelic Doctor has bequeathed to us, and We, therefore, declare that all the ordinances of Our Predecessor on this subject continue fully in force. . . . Further, We admonish Professors to bear well in mind that they cannot set aside S. Thomas, especially in metaphysical questions, without grave disadvantage.

The Angelic Doctor is a Scholastic appellation for Thomas Aquinas.

From all this Duhem had largely stood apart, and events will show him moving yet further away from it. Even in his historical work to date, unusual for its time in that Aristotle is taken seriously, there is no trace of the work on mediaeval science that was to be expected of a historically-minded Catholic scientist in that environment, and was in the end to do more than anything else to perpetuate Duhem's fame. But, as will be seen, when he does get involved in mediaeval science, what he finds is not perhaps what the Pope had in mind. While *Pascendi* was insisting on the centrality of scholastic philosophy Duhem was increasingly associated with a journal committed to opposing that philosophy. The best place to illustrate Duhem's developing distance from neo-Scholasticism is his association with the *Annales de Philosophie Chrétienne.*

5. The *Annales de Philosophie Chrétienne* and Open Apologetic

It is doubtful whether at the time Duhem was fully aware of the extent to which the years about 1905 represented a parting of the ways, and there are no radical changes in the journals he published in or in his relationships with their editors and owners. He continued to contribute to the *Revue des Questions Scientifiques*, then carrying his long-running *Origines de la Statique*, and which was to carry much else of his in the following years. He also contributed to the more determinedly neo-Scholastic *Revue de Philosophie*, which he seems to have helped his Brussels victim Jean Bulliot to found in 1900.[26] It had been launched with Duhem's *Le Mixte et la Combinaison Chimique* and, in these very years, was carrying his *Théorie Physique*. It too was to carry many more of his articles in the future. But that journal was to become the organ of the Société St. Thomas d'Aquin,[27] to which Duhem had long ago conceived a strong dislike, and to carry extended articles critical of his approach to scientific method, as well as papers by the young Jacques Maritain.

But 1905 was also the year in which he agreed to collaborate with the *Annales de Philosophie Chrétienne*, acquired by Blondel in that year on the death of its editor Charles Denis.[28] It had been founded in 1831 by Augustin Bonnetty (1798–1879), a disciple of Félicité de Lamennais (1782–1854), who gave it an orientation towards apologetics it retained right up to its disappearance in 1913.[29] His programme was natural theology of a type analogous to the contemporary English *Bridgwater Treatises*, but he cast his net wider, including things like law in his purview. In its early years it followed a policy of giving its natural theology out in little titbits drawn from the various fields of knowledge, but it soon settled down as a very long running conventional magazine. Under the editorship of Charles Denis from 1895, it changed its orientation away from science and towards the philosophical and biblical issues that had come to the fore in French Catholic discussion, a response, perhaps, to the rise of neo-Scholasticism. At that time it declared its orientation on its back cover as:

> At once classical, critical and apologetic, organ and tribune of all those within Catholicism preoccupied with religious, exegetical and scientific problems as they relate to the traditional theology and history of the Church.

It also announced, without, tantalizingly, naming names:

> This periodical has, both in France and abroad, more than sixty collaborators: clerical, lay, and teachers in various universities.

In due course it became the official organ of the Société St. Thomas d'Aquin, though, as was to appear in an article in April 1905,[30] shortly before his death, Denis personally was hostile to the neo-Scholastic movement because of the difficulties it made for serious apologetic.

Denis's successor shared this hostility. After he died, there was some negotiation before it was announced that he had been succeeded by a committee, with, as "editorial secretary", Lucien Laberthonnière, the Oratorian priest who wrote Denis's obituary.[31] The kingpin, however, was the new owner, Duhem's friend and contemporary at the École Normal Supérieure, Maurice Blondel. A list of "principal collaborators" on its back cover included some of the most significant

Catholic intellectuals of the time such as Friedrich von Hügel. In addition to Duhem and not a few of his friends, there were the names of several who like George Tyrell were to suffer in the Modernist crisis (not Alfred Loisy who had already been condemned by Rome). Their masthead bore a quotation from St. Augustine that put any suggestion of ultimate certainty out of court:

> Let us search then like those who must find, and find like those who must continue to search, for it is written: "the man who has reached the end is only beginning". (Ecclus 18.7)

In the first issue, the extended leading article 'Notre Programme'[32] made the whole thing explicit. It seems that it was drafted by Laberthonnière and the drafts circulated in the Blondel circle for criticism.[33] The following passage can serve as a commentary on the Augustine quote:[34]

> By the rational investigation it [the *Annales*] means to devote itself to it proves two things equal in importance and wrongly subjected to attempts to separate them: in the first place philosophical thought . . . is only complete and whole when searching for faith, and for having found it is only the more lively, hard working and free; secondly faith . . . is only complete and whole when searching for understanding and light, and for having found it, it in its turn is only the more profound, the more wide and the more sure.

This brief statement of the characteristically Blondelian philosophical position can be supplemented by another that spells out its implications:[35]

> Since we consider that the essential aim of philosophy is not to satisfy itself and stick fast in its ideas, but to promote life and thought the one by the other, and since we are convinced that in the real conditions of existence thought and life find their balance only in Christianity, it is naturally towards an apologetic that we will aim to complete our efforts and our researches.

> But it might be guessed that all that will matter for us is a lawyer's pleading, a mere defensive apologetic that, staying within a system of concepts and explanations, aims only to remain there. Truth does not conform to our narrownesses, and is stronger than all our presuppositions. . . . God protect us from trying to set ourselves up as a closed school! . . . In these conditions and despite the aim it pursues, the fact of writing in the *Annales* does not necessarily imply that one is Christian or Catholic, or even that at the present time one feels drawn towards becoming one. It implies only that one searches, and that one has the true desire to bring a solution to the problems and conflicts.

The article then continued with this crucial sentence: "and since we must only address competent readers, we will not hesitate, as occasion offers, in order to clarify *our reasons for belief*, to invite sincere minds to make known to us *their reasons for not believing*". Later, in response to the anti-Modernist encyclical *Pascendi* of 1907, they confirmed that they expected to have atheists in their readership,[36] as well they might when Catholics were beginning to find a welcome in such non-Catholic journals as the *Revue de Métaphysique et de Morale* (founded and edited by the Jews Xavier Léon and Léon Brunschvicg), which in its turn advertized itself in the *Annales*. It is hardly conceivable that a journal with such an open editorial policy would have tolerated contributors whose aim was to defend the Church at all costs. The issue that carried 'Notre Programme', the article proclaiming that policy, also carried the first instalment of 'Physique de Croyant', whose chief burden, as already noted, was, contrary to Abel Rey, the religious neutrality of Duhem's views on the philosophy of physics. The February issue carried Rey's brief, sensitive, and affectionate reply.[37]

The journal thus committed itself to an open non-authoritarian apologetic strategy: over the years it amply fulfilled that commitment. Much of it was written by Blondel himself, either under his own name or under one of his many pseudonyms. It lived dangerously on the frontiers of the faith, and in due course paid the price. As a result of the campaign by Blondel against Action Française discussed in the last chapter the *Annales* was denounced to Rome by its reactionary enemies and put on the *Index* in 1913 for reasons that doubtless included both its non-authoritarian stance and its hostility to neo-Scholasticism. For that too was not in doubt: Lucien Laberthonnière's *Le Réalisme Chrétien et l'Idéalisme grec* of 1904[38] had argued, in an interesting parallel to the later views of the Dutch Protestant Reyer Hooykaas,[39] that Christianity was radically incompatible with all Greek thought, and for that thesis it had been put on the *Index*. It was only later in his life that Blondel was to see any good whatever in the philosophy of Aquinas. In the deepening crisis of Catholicism of those years, to be associated with it was, whether Duhem liked it or not, and the evidence is that he did not, to be associated with one party in a passionate conflict. But that is to anticipate. His association with the *Annales* had a history, and something now needs to be said about that history.

6. The Passionate Anti-Scholastic

In 1896 Maurice Blondel was late in sending his New Year's greeting to his friend. His barely legible post card, dated 11 January, consists mainly of family civilities. The meat is in the middle. Does Duhem know that there's a certain Abbé Denis who is a man of good will and has been after Blondel for an article, in response to which Blondel has done him the charity of a polemical article, which Denis has published? Apparently Denis had the ardent desire to have a few pages of Duhem's prose. If Duhem consented, Blondel thought, the *Annales* would no longer be inept and the level would be raised. Blondel also asked if Duhem had seen what *le Monde* said of him. An extended quotation from Duhem's prompt reply[40] gives a clear view of his attitude of mind:

> The Abbé Denis must be a good guy, and the Société St. Thomas d'Aquin, of which his journal is the organ, also no doubt also contains good guys, but it also contains beings swollen with vanity—Count Domet de Vorges for example—as well as dirty poisonous beasts like the individual who hides in the *Annales*, and also hid in *Le Monde* under the name "a congressist'. I have no desire to mix my prose with people of that species, who think themselves authorized to tell lies because they wear soutanes.
>
> For the rest I will confess to you that people of this kind have left me disgusted with the Catholic world—I don't say with Catholics, which isn't the same thing—beyond all expression. The Lille Catholic faculty had already given me the measure of the sincerity reigning in that world, but the Brussels congress completed my education. Scribes and Pharisees, Hypocrites.
>
> My quite fixed intention is never to commit myself with those people. Search for the truth, and when I've found a particle of it, throw the news to the four winds, and then let the crows caw! So, the poor Abbé Denis won't be getting my prose.
>
> I havn't seen the articles in the *Monde* you mention and regret it, particularly if they were cutting me up: it amuses me to see the Catholics throwing themselves at me. If they were saying good about me I would be hurt.

The neo-Scholastics are not Duhem's only targets—liberals also get their share of Duhem's flak in the continuation of this letter to an intimate friend—but they are prominent. The reference to the Lille Catholic faculty is in direct contradiction to the story of friendly relations reported by André Chevrillon[41] for Duhem's daughter's biography of him. The incident at the Brussels congress two years earlier has already been referred to earlier in this chapter. It was something of a *cause célèbre*. There are accounts of it in greater or less detail in all reports of the Congress and books about the Catholic intellectual life of the time.[42] Duhem seems to have regarded the type of philosophy practised by the clerical philosophers of his time as altogether fraudulent. It is not difficult to see the reasons for Duhem's irritation: their quality is low. One wonders why he bothered, but he was a Catholic who had to live with that stuff while I am not. Duhem was so unimpressed by some neo-Scholastic groupings that at that date he was going to have nothing to do with any journal that harboured them. He might happily publish in the *Revue des Questions Scientifiques*, and later in the *Revue de Philosophie*, but he was not going to share covers with the Société de Saint Thomas d'Aquin.

In 1905, however, he had changed his line. The Société de St. Thomas d'Aquin had now transferred from the *Annnales* to the *Revue de Philosophie* and the *Annales* was now in the hands of Blondel. Blondel's letter has been lost, but Duhem's reply,[43] less bitter than the last one, conveys something of Duhem's despair in the deepening crisis:

> I ask for nothing better than to give pleasure to Fr. Laberthonnière. as also to Fr. Bulliot, as also to a crowd of nice guys whom I've nothing against except that they shoot each other too much instead of uniting against the common enemy. So it is understood that if one of these days I write something suitable for the *Annales*, I will send it in, without thereby breaking with the *Revue de Philosophie*.

and, as good as his word, he give the *Annales* his 'Physique de Croyant', the piece discussed above whose first instalment appeared alongside 'Notre Programme' in the issue marking the formal inauguration of the new regime. It is hard to conceive of a more suitable article for the purpose. The Blondelian flavour of its strategy does not seem to have been previously remarked on: a detailed analysis of the logic of physics purports to cut off all possibility of a science-based natural theology and of any science-based attacks on Catholic belief, while leaving open the possibility that the actions of the Church may have been intellectually beneficial. Nonetheless it is unmistakable, and despite the differences between Blondel and Duhem that remained, it was just the thing to inaugurate the Blondelian programme for the journal. In due course the *Annales* would carry *To Save the Phenomena*, and another piece was about to be published in it when it was put on the *Index*.[44]

Shoot at each other the good Catholic guys continued to do, and it would have been difficult for Duhem not to have had his sympathies. In deep sympathy with his student friend Blondel, he had long, for reasons that will become clear in the next chapter, had his intellectual preferences, which were not neo-Scholastic. I have already discussed the evidence of Duhem's sympathy for the journal and its policies in these later years. The apparently grudging collaboration offered in Duhem's letter of 1905 was real and deep, reflecting the opposition he shared with his friends to the neo-Scholastic philosophy and authoritarian attitudes then encouraged by the Catholic authorities.

It is possible to speculate on the roots of these attitudes and what encouraged them. No doubt they must have began at school, with exposure to third-rate manuals and much of the like, but whatever their roots they were certainly reinforced later in his career. Central to Blondel's assault on Action Française and its allies was a critique of what he called the "monophorisme" of the Scholastics, their view that only their philosophy was capable of conveying the gospel. The persecution of Laberthonnière (after 1913 forbidden to publish anything right to the end of his life)[45] and the perpetual difficulties of Blondel can hardly have endeared him to the movement in whose name these atrocities were committed. To judge by extant correspondence, his Bordeaux friend and colleague Albert Dufourcq may also have encouraged his attitudes.

The most remarkable thing about them, however, is the passion with which they were pursued and held. It is tempting to apply to Duhem the famous saying of David Hume, "Reason is and ever ought to be the slave of the passions":[46] it is the sheer passion that comes to the fore in any consideration of historical work of Duhem's later years culminating in the *Système du Monde*. In a later chapter I discuss the evidence for and the meaning of his resistance to mediaeval science before 1904, but none of these are likely to explain the passion with which

Duhem subsequently dived into that area: they would only help if the passion was already there.

It has to be remembered that we are dealing with a man of extreme independence of mind. Never afraid of attacking the great and the powerful it is truly reported of him that "All his life he put into practice the adage *Amicus Plato, magis amica veritas*"[47] (Plato is a friend but truth a greater). His friends seem to have learned to allow for his characteristic violence of style, but it was a constant feature. Never a believer in intellectual compromises, he always took everything to extremes. Among the results of that attitude are the celebrated attack on Berthelot—natural enough, we might say, since Berthelot was a republican positivist and Duhem was not, but Duhem's readers do not seem to have realized that it was possible for Duhem to pick on Catholic victims as well as positivists. I illustrate this with another quotation from the 1913 letter to Blondel cited above: "They have got into the habit of a sort of verbosity, a play with formulas borrowed from St. Thomas or others, with which they think they answer everything when they are answering nothing at all." Duhem's opinion of their historical competence wasn't high either: "They tell us a heap of things with no relation to historical reality."[48]

The collaboration with the *Annales* provides, then, one perspective from which to consider the effects of these preferences, one of which was to carry him away from neo-Scholasticism and all its works, to bring about a decisive turn away from the intellectual programme the ecclesiastical authorities were so urgently promoting. Surely, it will be said, the counsels of prudence Duhem was in due course to invoke over Galileo's head in *To Save the Phenomena*[49] should have applied here too, and encouraged in him a willingness to compromise? It seems not. Compromise, of course, was not in Duhem's nature, but the lengths he went in the other direction were extreme: even his Bordeaux crony Albert Dufourcq thought it was going a bit far to condemn Aquinas as a well-meaning philosophical incompetent whose famed 'synthesis' of Christianity and Aristotle was nothing of the sort, merely incoherent![50] I have already hinted at some of the passions that took Duhem this way, but more must now be said about the sources of Duhem's ideas, and the intellectual tradition that fed them. For whatever exposure Duhem may have had to scholastic manuals at the Collège Stanislas, it is abundantly clear that his prime inspiration came from elsewhere, from Pascal. It was in Pascal that he found the epistemological resources for his apologetic enterprise of drawing the anti-religious sting from positivist philosophy.

IV

Inspiration from Pascal

1. A Pascalian in the Pascal Revival

"He was once . . . labelled a Kantian . . . [but] . . . no, Duhem depends not on Kant, but on Pascal, on the Pascal he is always citing, whose *Pensées* he knows by heart."[1] The comment comes from Duhem's lifelong friend, the mathematician Émile Picard, now permanent secretary of the Académie des Sciences; the occasion was a commemorative address to the Académie; Kantianism was an accusation often made by neo-Scholastics to damn the philosophies they liked least, the context on this occasion being Duhem's well-publicized resistance to Scholastic natural theology at the 1895 Brussels congress. Picard's strong claim is confirmed by the equally strong claim of Duhem's sometime Bordeaux colleague and Pascal scholar Fortunat Strowski, near the end of a somewhat informal article[2] apparently brought forth by the tercentenary of Pascal's birth: "At Bordeaux I knew well a great scientist who had reflected more than anyone on the history of the sciences, on the methods of science, and on physical theory: Pierre Duhem. He never stopped appealing to the example of Pascal, never gave a lecture, never wrote a chapter, without citing the *Pensées*; he it was who gave me my knowledge of and taste for it." Duhem's Pascalian interests were in any case known: on one occasion Duhem's friend and correspondent Jules Thirion actually wrote Duhem asking for help in tracing a Pascal reference used by Strowski;[3] he had been publishing on

Pascal's experiments in hydrostatics and the weight of air;[4] his is the preface to Albert Maire's bibliography of Pascal's scientific work,[5] reproduced in the appropriate volume of Maire's later comprehensive Pascal bibliography;[6] at all stages of his career Pascal citations were an obvious and unmistakable feature of his work. His continued use of the nineteenth-century Havet edition of the *Pensées*[7] long after it should have been superseded by more modern editions, and his increasing habit of not marking the elisions in his citations, confirm the implication of Picard's statement, that he cited from memory. The Ariadne's thread to Duhem's work postulated in Chapter I is to be sought neither in Thomas nor in his modern followers, but in Pascal. The element of exaggeration in claims like Strowski's and Picard's concerning the frequency of Pascal citations in Duhem's work should not be allowed to obscure the evidence of the impression Duhem's Pascalian orientation made on his colleagues and associates.

Duhem's interest in Pascal was by no means unusual: even by the standards of French culture, for which Pascal's writings represent one of the great literary monuments of the seventeenth century, his time was one of particularly intense interest in Pascal, but it does seem that even in that time, Pascal found in Duhem a reader who by training and orientation was specially suited to appreciate the range of Pascal's interests. In what follows I develop an account of the features of Pascal's work relevant to an understanding of Duhem.

2. Pascal and His Work

Astonishing in the breadth of their range, Pascal's interests ran all the way from transport via mathematics to theology.[8] In a man who was not 40 when he died in 1662, his achievements are truly astonishing. Over a hundred years before Shillibeer, he devised and operated briefly an omnibus service, 'les carosses à cinq sols'. Over 150 years before Babbage, he designed, built, and sold working calculating machines that used the ten-toothed wheel for counting tens, a principle not superseded till the advent of electronic techniques after the Second World War. As a mathematician, he earned the respect of contemporaries such as Fermat and Huygens for his work on the cycloid, and he was a pioneer of the mathematical theory of probability. In physics a significant mechanist critic of Descartes, he had a decisive influence on the development of ideas of fluid pressure. Holding, as we do now, that effects previously attributed to a supposed abhorrence of a vacuum on the part of nature were caused by nothing more mysterious than the weight of air, he organised his brother-in-law and second cousin Florin Périer to check the height of Torricellian mercury barometers at different altitudes on the Puy-de-Dôme in the Auvergne.[9] Whoever first suggested this experiment (possibly Mersenne[10]) its successful outcome did more than anything else to establish the modern view of the matter. To that celebrated experiment may be attached the correspondence with the Jesuit Père Estienne Noël[11] in which Pascal displays his methodological acuity in criticizing both Aristotelian and Cartesian natural philosophy. For Pascal, as later for Duhem, Cartesian subtle fluids were no improvement on Aristotelian occult qualities. His *Préface sur le Traité du Vide*,[12] that like much else in his work remained unpublished, deplores excessive respect for tradition: Pascal considers that in science this is as pernicious as a lack of respect for tradition in theology. With these writings go the so-called *Opuscules*, of which the most important are perhaps the *Esprit Géométrique* and *Art de Persuader*.[13] In addition to the discussion of geometrical method that was used (and mis-used) by Arnauld and Nicole in their *Logique ou l'Art de Penser*, the so-called *Port Royal Logic*, these also contain a discussion of the infinitely great and infinitely small that was taken up again in the *Pensées*.

To the bus operator, physicist, and mathematician, has to

be added the theological pamphleteer and theologian. Two conversions took the Pascals as a family and Blaise in particular away from an earlier worldly phase, and led them to turn their backs on their former worldly interests and concerns and become fervently pious members of the so-called Jansenist circle. This embattled group, centred on the monastery of Port Royal des Champs, defended through thick and thin the presentation of the philosophy and theology of St. Augustine given in the *Augustinus* of Bishop Cornelius Jansen of Ypres (1585–1638), and besides Pascal its most famous member is probably the theologian and philosopher Antoine Arnauld (1612–94), correspondent or opponent of most of the major philosophers of his age. The *Lettres Provinciales*[14] and the *Pensées* not only remain classics of French literature but retain an interest for a much wider audience than the purely literary. The former shows Pascal to be an able pupil of his Jansenist mentors Antoine Arnauld and Pierre Nicole (1625–95). Though Pascal may have failed in his original object of forestalling Arnauld's expulsion from the Sorbonne, the campaign against Jesuit moral theology and casuistry seems to have done permanent damage to the reputation of the order, with effects that persist to this day in the pejorative connotations of the adjectives 'casuistical' and 'jesuitical'. It has not always been realized that the *Pensées* are not a collection of random thoughts on religion, but the unfinished result of Pascal's attempt to produce an apology for the Christian religion, and that in working on it he had to evolve his answer to the sceptical doubts about the possibility of knowledge widespread in the seventeenth century.

3. The Interpretation of Pascal

Though Duhem's interests were not quite as wide as Pascal's, the overlap between their scientific and religious concerns was considerable: with that and the lifelong ill-health he also shared, he was, *prima facie,* well-placed to appreciate Pascal, but which Pascal? Even the most sympathetic of Pascal's readers find more for them in some parts of his work than in others, and put different interpretations on it. They may, for example, prefer the mathematics to the physics, or *vice versa,* and between the scientific parts of his work and the theological, preferences can and do vary, many writers concentrating on the science to the exclusion of the theology, or the other way about. Even when the appropriate texts have been selected, the task of deciding on the construction to be put on them is frequently difficult in the extreme. In Pascal's work texts are often particularly difficult to date and place in context: the collection of notes, known to the world as the *Pensées* is only the best-known and most extreme example of a perpetual problem for the student of Pascal's writings that affects both the scientific and the theological writings. In the same case are prefaces for projected works on physics and geometry not completed or published and containing material of crucial importance for an understanding of Pascal's theory of knowledge, for a theory of knowledge Pascal certainly had.

The interpretation problems arise because of the problem of matching these dateless texts with the known evolution of Pascal's career from his early mathematical and scientific interests to his later theological concerns. Can the fragmentary prefaces of *Opuscules* be used as evidence for Pascal's later position, when he was writing the *Pensées,* or are they too different in date to be any help? Pascal's associates at Port Royal encouraged a picture of a man who had dropped all his previous scientific concerns for the sake of his immortal soul, and to the celebrated Dutch physicist Christian Huygens, he was in that phase "dead for geometry".[15] Modern writers are not so inclined to draw such sharp boundaries between Pascal's various concerns, and are apt to recall that in his last years, when he was certainly already deeply involved in planning his

Apology for the Christian Religion, he stage-managed a competition among mathematicians to solve the mathematics of the cycloid, a competition in which he triumphantly declared himself the winner. An apology, any apology, for the Christian religion has to attend to epistemology, the theory of knowledge, if it is to have any hope of credibility, and the *Opuscules* set out the foundations of one such, one that cannot be irrelevant to the projected *Apology* of which the *Pensées* is all that remains.

Problems of interpretation, problems of context, problems of text: all these come together in dealing with the *Pensées,* the document by which Pascal is most widely known, and which has attracted more scholarly effort than almost any other save the Bible, for it raises in an acute form the question of just what the reader of Pascal is looking for: spiritual sustenance, Biblical scholarship, moral teaching, speculative theology, philosophy, metaphysics, literary style, or what? All these things can be found in the text.

The *Apology for the Christian Religion* Pascal had been working on an at the end of his life was to deal, in his way, with a problem felt as urgent by many of his contemporaries. He had worked out a plan for it and, in a lecture delivered at Port Royal, expounded this plan in outline.[16] He had also been accumulating notes to serve him in working out this plan, and even begun classifying these notes before illness and death finally stopped work. When he died, his Port Royal executors considered publishing his files as they stood, but decided against that course for reasons they did not state but which can easily be guessed at: embattled in the defence of their position under near-perpetual persecution, they were enjoying a respite, which they were in no mind to jeopardize. The safe thing was a collection of pious words from a man, who for all his combativeness, was of known sanctity, a collection that left out anything that might recall the philosophical and religious controversies of their time. Thereby they did two things:

they established the tradition of publishing thematic arrangements from a collection they presented as random; and by ransacking the papers for the material that suited their purpose, they left them in a state of increasing disorder, a position in which further deterioration was only finally prevented by the preparation of a paste-up of the fragments that resulted. This so-called *Recueil Original* has been fundamental for all Pascal editions ever since.

Pascal had the gift of striking phrase: his frequently enigmatic utterances fascinated successive generations of readers and editors who came to study what became a monument of French literature from the *Age Classique,* readers and editors who were in no position to appreciate the internal structure of Pascal's thought, in no position both because of the fragmentary nature of the text they inherited, and because they were increasingly remote from the age in which Pascal lived. Few if any were committed Christians, and even fewer shared Pascal's theological and philosophical outlook—Voltaire, Havet, Brunschvicg, atheists or deists all, got to work on the *Pensées.* Perhaps the key event in the story was Victor Cousin's discovery in the second quarter of the nineteenth century of the textual inadequacy of the available editions when compared with the manuscript. As a result successive editors such as Prosper Faugère, Ernest Havet and Léon Brunschvicg set to work to produce editions meeting the higher standards called for by such an important text. Each, though, saw himself as editing a collection of *fragments,* whose intended rôle in Pascal's overall scheme was a matter of speculation. There were two possible courses of action: to attempt a speculative reconstruction of Pascal's scheme on the basis of information from such sources as Filleau de la Chaise's account of Pascal's Port Royal lecture, or merely to arrange the material according to the editor's conception of the themes. The first course was taken by Faugère and Havet in the nineteenth century and Jacques Chevalier in the twentieth; the second, probably inevitable in view of the unavoidable arbitrariness involved in the other course, given the sparseness of the evidence available, was taken in the early years of this century by Léon Brunschvicg, whose edition is probably still the most widely used: at least arrangement by theme means, if it is well done, that the reader can easily look Pascal up on the topic that concerns him or her. But the chaos resulting from the multiplicity of editors and editorial principles can be readily appreciated: every editor divided up his material into a different system of fragments and provided his own reference system for his fragments, inevitably incompatible with all others. However, a second major event promises a way out of this confusion.

During the Second World War Louis Lafuma realised the importance of the surviving copies of Pascal's original manuscript, copies made with considerable care before the files were cut up by the Port Royal editors, and revealing a partial classification of the material. Lafuma believed, and later scholars have agreed with him, that these copies, whose existence is referred to in the preface to the Port Royal edition of *Pensées,* were prepared to assist the preparation and checking of Pascal's material for that edition, and that the classification they revealed had to be Pascal's own, and respected as such, incomplete as it was. They differ in the order of the material, but they do not differ in the classification they reveal. It seemed to Lafuma that any edition of the *Pensées* ought to respect Pascal's own ordering and he based his on one of these two copies. But that claim still left two questions: 'Which copy?' and 'What about the material Pascal had not got round to classifying?' Lafuma chose the so-called First Copy, and Sellier the Second, on the basis of arguments that the latter is relatively more free from corruptions and was the one used by Pascal's sister for checking. Lafuma actually prepared two editions: the first time he attempted to fit the unclassified papers into the framework provided by Pascal's own classification; the second time he abandoned even that amount of editorial intervention and aimed at an edition in which there was no editorial screen between Pascal and his modern reader. Philippe Sellier has gone even further in the same direction.

If modern editors have tried to keep editorial intervention to a minimum, their nineteenth-century predecessors pursued no such ideal. Mostly, as remarked above, atheists or free-thinkers of various sorts, for them Pascal was an intellectual scandal, a scandal because of the potential persuasiveness of his advocacy of a religion they opposed, and a scandal because of what they interpreted as its radical scepticism, an interpretation hard to resist as long as the material could be treated as no more than a collection of fragments with no overall plan. Typical, and perhaps most influential, was Ernest Havet, a teacher of ancient literature who was to conclude his career with a *Le Christianisme et ses Origines* that had the *hubris,* despite Havet's ignorance of Hebrew, to argue, via some radical redatings of the Old Testament prophets, that everything worthwhile in Christianity was derived from Greek philosophy. Havet's edition of the *Pensées* was accompanied by a 54-page *Étude* and a massive critical apparatus that in places argues with Pascal point by point, so great was Havet's determination to resist Pascal's 'scepticism'. Any user of this edition would have had to confront this interpretation and this apparatus, and if he disagreed would have been forced to pay close attention to the text, and into long meditation on it before he could hope to fight his way out and establish his own position. Duhem did use this edition, and never refers to any other. He did reach an interpretation very different from that of Havet. To achieve it he would have been forced to read Pascal from the epistemological point of view, at least as much as a spiritual classic. As will be seen, this is just what we find.

4. Scepticism and Infinite Regress

I have referred to the scandal caused by Pascal's perceived scepticism. Readers will be familiar with scepticism as a term for anti-religious attitudes. In this context, however, it has a wider connotation: doubts about knowledge claims of any kind. Pascal was perceived to be entertaining arguments that cast doubt on the reliability of science as well as religion. For the reasons indicated in Chapter II, this was a scandal for positivist atheists, including Havet, for whom science offered a paradigm of reliable knowledge; at the same time it was also, for the reasons indicated in Chapter III, a scandal for scholastically-minded Catholic apologists who needed a reliable science to support their natural theology, and for whom, as with Vicaire, scepticism about any branch of knowledge implied religious scepticism as well. But since, scandalous or no, this was the issue at the heart of the *success* of the *Pensées,* and central also to the use Duhem made of Pascal, something needs to be said about it here. The issues involved will perhaps emerge most clearly if discussion is focussed initially, not on the *Pensées* but on that much misunderstood and traduced text, the *Esprit Géométrique.*

Originally intended as a preface for a textbook of geometry that, if it was ever completed, has now been lost, this text addresses the question of the method of geometry, the discipline commonly regarded, then as now, as the most rigorous of all. To focus the discussion, Pascal characteristically introduces a model of extreme rigour, the "true method" of constructing "the most excellent demonstrations if that were achievable":[17] that method in which no proposition is advanced without first being proved and no term introduced without first being defined, an ideal no reader should have any trouble understanding—for have we not all been asked in controversy to prove questionable claims and define doubtful terms? Certainly, says Pascal, it is a beautiful method, but quite impossible, impossible because any proof must suppose or take for granted things that have not been proved, and any definition must start from or take for granted things that have not been defined. The point was certainly not original to Pascal, but since it is unfamiliar and somewhat technical, I will now attempt to expand on it and bring out its implications.

If I want to prove that the reason a car will not start is that the battery is a dud, it is normally enough to replace the battery in question with a new one, and establish that it now starts without any difficulty. If we accept that the car starts with the new battery, and that it does not with the old, we would regard ourselves as having *proved* that the battery was the problem: two propositions are combined into a *proof* of the third. It is always thus: proofs must always start with other propositions for the moment at least taken for granted. But, and this is Pascal's point, if we want perfect rigour, we have to prove these too. After all, the battery may be perfectly OK, and merely suffering from greasy contacts, or—if we are seriously interested in rigour we have to consider every possibility, no matter how remote—someone may have jinxed it and appropriate magic spells would then be needed to put things right. It is the same with definitions. To complete in perfect rigour Aristotle's celebrated definition of man as a political animal we need definitions of 'animal' and 'political'. If the definitions of these involve 'living thing' and 'government', then definitions of these will be required in turn. At this point the reader who has not been philosophically trained, as well as many a reader who has been so trained, is likely to become impatient: 'Surely', he will say, 'there are always available some things that do not need proof or definition, and we can start from these?'. Just so. The problem is to identify these; they tend to vanish under criticism, and the certainties of one age often disappear the next.

The difficulty under discussion is technically known as *infinite regress*, because when complete rigour is the goal, there seems no stopping the demand for proofs of everything used in proofs, and definitions of everything used in definitions, unto infinity. It may be thought that this discussion too takes something for granted: the ideal of complete rigour, without which it could not even get off the ground. Without this supposition, it may be said, Pascal would have been unable to make the point that it was an impossible ideal, beautiful but impossible. Without it also, Duhem's own arguments could never have got off the ground: in fact this supposition will be fundamental for the argument of this chapter and the next; as a supposition it was hard in any case to avoid, given a serious attempt to state just what proof consisted in, what was it exactly that distinguished good arguments from bad, hard to avoid even for the modern probabilist who, no doubt rightly, seeks an alternative model for what he considers a good argument, for it is relative to a rigorous argument that the properties of probabilistic arguments are investigated. In fact, though, infinite regress will catch the probabilist anyway: a probable argument has to start from things that are at least probably true but have not been even probably proved, and once again there is no stopping the infinite regress. It seems that with infinite regress we have encountered an 'engine of war' capable of casting doubt on any piece of knowledge whatever. For if it can force us to recognize that even geometry rests on nothing more than unproven suppositions, where on earth or in heaven are we to look for reliable knowledge, the ideal that has played such a large part in Western culture?

Thanks to the work of Richard Popkin[18] and the late Charles Schmitt,[19] the scholarly world is now familiar with the crucial rôle played by sceptical doubts of this kind in the culture of the sixteenth and seventeenth centuries. In Popkin's account, Reformation debates about the rule of faith, be it Church or Bible, were in effect debates about the criteria for religious knowledge. As such they were irresolvable in principle, and led to doubts about the basis for all knowledge, and to a renewed interest in the writings of the ancient sceptics, especially the *Pyrrhonian* Sextus Empiricus. Sceptical doubts were publicised by the *Apologie de Raimond Sebond* in which Michel de Montaigne[20] 'defended' Sebond against charges of heresy on the sceptical grounds that there was no reliable basis for either Sebond's position or the opposite. Used extensively by Catholic apologists to cut the ground from under their Protestant opponents, it soon emerged that scepticism was a double-edged weapon, as dangerous to its user as to his intended victim, and that there was no sphere of knowledge safe from sceptical attack. Scepticism had let loose an 'engine of war' of unlimited destructive power, so that there was generated a 'Pyrrhonian crisis' in which overcoming the sceptic was seen as an increasingly urgent priority. The most famous of many unsuccessful attempts at ending this crisis by the introduction of a more soundly-based new philosophy was that of René Descartes. Observe again that the scepticism here at issue was not, until very late in the seventeenth century, directed against religious belief as such. Used in the first place, with no sign of a bad conscience, by one religious faction against another it supplied the essential context for the apologetic enterprise of Pascal *on behalf of religious belief*, as well as for commonly encountered so-called *fideist* positions, in which religious belief is held on faith alone without any attempt at supporting that faith with rational grounds.

5. Ways of Dealing with Infinite Regress

The reader of Pascal familiar with the writings of Richard Popkin has no difficulty in recognizing an obsession with Montaigne and Pyrrhonian scepticism everywhere in the *Pensées*. My concern now, however, is to consider how Pascal dealt with these sceptical doubts, for despite editors like Havet, Pascal did *not* hold that 'Pyrrhonianism' is the truth—a statement rather like 'All Cretans are liars' in the mouth of the *Cretan* Epimenides—if true it is false, and if false, true. To get a sense of what the options are, it is well to begin with Aristotle, almost certainly the originator of both the infinite regress argument and of the formal logic without which it could hardly have been formulated. At the close of the *Posterior Analytics,*[21] the work in which he used his logic to set out his view of true knowledge, and presented his model of deductive sciences resting on true principles, Aristotle considered the question of how these principles were to be obtained. It cannot, he says, be deduction, or demonstration, because that would simply transfer the problem to the principles on which the postulated demonstration rests, for *they* would now be the principles of the science in question. Aristotle's answer, *epagôgé*, often translated 'induction', seems to be a kind of intuitive process in which in the course of continuous immersion in experience, the principles of the science emerge from the contemplation of many instances of its objects. Hardly, you may say, a satisfactory answer, but infinite regress is going to make it hard to devise a better alternative free from like objections.

Aristotle's treatment of the problem may be compared with that of Leibniz, in a letter of April 1686 to his Paris sceptical friend and canon of Dijon, Simon Foucher, a letter also notable for its succinct summary of the metaphysical system of pre-established harmony first developed in the *Discourse on Metaphysics* drafted early in February that year.[22] Foucher had made something of a name for himself by his sceptical critique[23] of the *Recherche de la Vérité* of Nicolas Malebranche, though by the mid-1680s Foucher's criticisms must have been overshadowed by the heavy artillery of Antoine Arnauld.[24] He was always in difficulty getting publishers for his work. Anyway, in the spring of 1686 he had got together a parcel of his books for sending to Hanover, and when it arrived, the ever-eager Leibniz seems to have devoured them at a sitting. Leibniz, as is well known, usually liked to find himself in agreement with his correspondents, so that the significant issues are always where he chooses to express disagreement. Here, the issue he lighted on was Foucher's use of the term 'supposition' as something like a term of abuse. Leibniz objected: not only would rigorous avoidance of suppositions slow down progress in the sciences, a point he had already made to Foucher ten years earlier, but some suppositions had to be made if anything is to be established at all. Proofs could not go to infinity. Making suppositions to build rigorous demonstrations on, wrote Leibniz, was the true method of the geometers, instancing the work of Archimedes; and Foucher too, he claimed, was compelled to make suppositions in reasoning and discussion; otherwise he would be perpetually liable to say both one thing and its contrary. Ten years earlier he had offered this same correspondent the method of reduction to identities as the answer to the problem of scientific certainty: now all he offered were suppositions, though the suppositions he offered may cause surprise: the principle of contradiction and the principle that in every true proposition the predicate is included in the subject: surprise because both of these have a flavour of logic about them, the second being the principle the whole of Leibnizian metaphysics is commonly thought to rest on, and which certainly plays a large rôle in the *Discourse on Metaphysics*; but it remains the case that there is no way of proving these principles, they are simply supposed, and they would remain suppositions even if it were universally agreed that they were the only reasonable suppositions to make.

Pascal had his own way of dealing with the problem. In the *Esprit Géométrique*, or rather the *Art de Persuader* usually included in most editions of that document, this is not developed very far: and in a passage[25] quoted out of context by Arnauld and Nicole in the *Port Royal Logic*,[26] he contented himself with saying that you didn't attempt to prove what did not need proof, and set out a collection of rules to that effect. It is in the *Pensées* that the full implications of his position emerge, with his doctrine of the heart that has its reasons that reason does not know:[27]

> We know the truth not only by our reason but also by our heart. It is by the latter that we know the first principles and it is vain that reasoning, which has no part here, attempts to defeat them. The Pyrrhonians, whose only object is this, attempt it uselessly. We know we are not dreaming: however powerless we may be to prove it by reason, that powerlessness proves no more than the weakness of our reason, not, as they claim, the uncertainty of all our knowledge.

Pascal's readers may agree that it would be pointless to attempt to prove that we are not dreaming, but they may be less happy with the sequel:

> For the knowledge of the first principles—space, time, motion, numbers—is as secure as any that our reasonings give us, and it is on this knowledge of the heart and instinct that reason must lean and base the whole of its argument. The heart senses that there are three dimensions in space and that there are infinitely many numbers, whereupon reason demonstrates that there are no two square numbers such that one is twice the other. The principles are sensed, the propositions prove, and both with certainty though by different routes—and it is as useless and ridiculous for reason to ask of the heart the proofs of its first principles before accepting them, as it would be ridiculous for the heart to ask reason for a sense for all the propositions it demonstrates before accepting them.

A natural response to the type of argument just quoted from Pascal is anachronistic: we now have a widely accepted and fruitful physical theory, that of General Relativity, that asserts that space has more than three dimensions, since it makes use of a four-dimensional geometry of space-time. It is tempting, therefore, to regard as refuted the geometrical teachings of Pascal's heart. The alternative view of geometry to Pascal's would be to regard it as a completely arbitrary construction on freely chosen principles. It is not, however, clear that this latter view is any more free of serious problems, and in any case, this passage does at least make clear that, in intention at least, Pascal is no sceptic.

In its general form, this passage bears all the marks of being directed at the Descartes of the *Meditations*, as well as, no doubt, at those geometers of his time who, like Roberval, wanted to demonstrate Euclid's axioms, for it was Descartes who not only entertained the hypothesis that we were all dreaming, but who also, perhaps to avoid infinite regress, proposed that the steps in his "demonstrations" needed to be sensed if they were to be relied on. However, here, Pascal has other ends in view, an *Apology for the Christian Religion*:[28]

> Hence this powerlessness must serve only to humiliate reason—which would try to decide everything—but not to contest our certainty. As, if only reason were able to instruct us, it would please God that on the contrary we never needed it and knew everything by instinct and feeling but nature has refused us this bounty: on the contrary it has given us very little knowledge of this kind: all the rest can only be acquired by reasoning.
>
> That is why those whom God has given religion by the feeling of the heart are blessed and quite legitimately persuaded, but to those who have not got it, we cannot give it by reasoning, until God gives it to them by a feeling of the heart, without which faith is merely human, and useless for salvation.

Reason and sense thus play different rôles in different subject-areas, which, it will be seen, are separated from one another, separated by method as they are separated by subject-matter. Equally, it will be seen, Pascal has separated off faith from physics in just the manner we have seen in Duhem, ruling out equally the use of natural reasoning to defend Christianity and to attack it. I shall return to this feature of Pascal's thinking in the next chapter; in the meantime it is enough to note that despite its author's expressed intentions it remains vulnerable to suspicions of scepticism. But in an age dominated by the pretensions of the positivist advocates of the universal competence of 'science', the very suspicion of scepticism attached to it could prove attractive to those inclined to oppose such positivist pretensions, a reaction which may be compared to the way in which the rejection of the excessive claims of the advocates of 'science' has led some in our own age to reject not just science, but every rational ideal, and to embrace fringe sciences of the most shaky kinds. The latter years of the nineteenth century saw an enormous vogue for Pascal that continued into the twentieth. As D. M. Eastwood has shown,[29] few French intellectuals of the time remained unaffected by it, not even the Comtian Maurras. Least unaffected were Duhem and his friends Blondel and Laberthonnière.

It is not my concern here to discuss the causes and follow the course of the literary success of Pascal's *Pensées* in late nineteenth century France: Eastwood's *Revival of Pascal* is enough for my purposes. The essential is to realise that, though the literary effects are visible enough the effects were not purely literary. They can be seen, for example, in the widespread predilection for dialectical modes of presentation and paradoxical expression, traits with important implications for interpreters of Duhem, where they go along with his natural tendency to extreme statement: in a manner highly reminiscent of Pascal, Duhem is continually stating positions which will be abandoned at later stages of his argument. But, surprising as these may be to those not used to regarding Pascal as an important figure in the history of epistemology, the effects that most concern me here are epistemological, in the theory of knowledge—not the least of the merits of Eastwood's book is the prominence she gives these—and it seems appropriate to say something now about their rôle in the culture of the time.

The prevalence of scientistic anti-religious attitudes in nineteenth-century France—and indeed in other countries—has been the theme of earlier chapters. I have commented above on the image of science they propagated: that of superior rigour, purportedly superior to that of both religious and literary writings. Such an image inevitably pushed into the foreground the epistemological question of the nature of scientific rigour. For those who shared these attitudes, Pascal was, as I have already remarked, a scandal, a scandal not only because of his combination of science with religious commitment, but a scandal also because he seemed to countenance scepticism; for that, in apparent ignorance of genuine scepticism, was how they read his critique of scientific rigour. But, when the time came for a philosophical and literary revolt against this scientistic climate, this very critique of scientific rigour was what most commended Pascal to many literary and other readers. Those readers who in Britain and Germany might have found sustenance in literary romanticism, came in France to be accommodated by Pascal instead. For in Pascal they found arguments that directly undermined the scientistic positions they found so repellent, arguments that purported to show that the heart, not reason, was the basis of all knowledge. It was of course all too easy for such readers to concentrate on the heart to the exclusion of reason, and to develop an epistemology that, as Duhem would put it, gave reason, instead of too large a rôle, no rôle at all, but excesses like this are usual in such reactions. It is enough for the present that the scientistic climate of the nineteenth century had focussed attention on epistemology, and it was epistemological inspiration that the Pascalians of the late nineteenth-century Pascal revival found in the *Pensées*. No doubt there were other things in Pascal, and no doubt readers like Duhem, steeped in Pascal since childhood, found these too, but epistemology was in the forefront of their concerns, as it most certainly was in the forefront of Duhem's.

V

The Appropriation of Pascal

1. *Géométrie* and *Finesse*

Pascal, then, offered much to a man like Duhem. The question now is 'What in fact did Duhem take from Pascal?', and the answer is surely 'A great deal'. Of that the most famous example is surely the distinction between *géométrie* and *finesse,* between geometrical and intuitive minds, that Pascal sets out in two famous fragments of the *Pensées,* using the terms *géométrie* for pure deductive reason, and *finesse* for intuitive judgement, those mental abilities that escape deductive reason.[1] Developing the argument of his 'École Anglaise' of ten years earlier, Duhem used it, as is well known, for a double purpose: on the one hand to attack the incoherences of 'English' (actually mainly Scottish) physicists who substituted the construction of models for careful logical deduction; on the other to show that there could be no grounds in logic for the demand that his *common sense* compelled him to make, that physical theories should be coherent. In developing his case Duhem claimed to discover a predominance among the 'English' of broad and weak minds with more than their share of *finesse,* compared with an opposite predominance on the Continent of strong and narrow minds favouring a more rigorous deductive style of physics. This cultural contrast is illustrated by the parallel development of the English novel, instanced by the ability of Dickens (whose work Duhem knew well and loved[2]) to hold together a multitude of different characters and story lines without losing any of them. At this point, Duhem saw Hertz's treatment of Maxwell, against which he protested, as an extension of 'English' physics away from its home territory.[3] To say with Hertz that Maxwell's theory *was* Maxwell's equations was to make a model of it, just as much a model as one of Kelvin's constructions with wheels and pulleys.

To many readers, despite Duhem's insistence that he was not in any way casting doubt on the greatness of physicists like Maxwell and Kelvin, material of this sort must seem no more than a chauvinistic parlour game, hardly serious epistemology, but first impressions would be mistaken. Ten years later, in very different circumstances, Duhem was to make much more extensive and systematic use of Pascal, and do so in a way which does, I believe, put his entire thinking into a new perspective, confirming all the intuitions of his friends and contemporaries about Pascal's importance for him. Along with *La Chimie, Est-elle une Science Française?* (in reply to Ostwald[4]) and 'Science allemande et vertus allemandes'[5] the *Science Allemande* of 1915 is one of Duhem's contributions to the flourishing war literature of the time. It consists of four lectures to an audience of Bordeaux Catholic students, circumstances which largely preclude precise technical argument, plus an article on the same theme reprinted from the *Revue des Deux Mondes*.[6] The changes in perspective are startling: instead of being blamed for imitating 'English' model-making, Hertz is now attacked for that excessive confidence in deductive reason that enabled him, ignoring the promptings of *finesse,* to choose any arbitrary set of hypotheses for his theory despite their irreconcilability with experience. Hertz's behaviour here is compared with Hegel's derivation of his metaphysics from the absurd (according to Duhem) principle of the identity of opposites. But these differences are merely of perspective and, despite Maiocchi's claim to the contrary,[7] the imprecisions of popular exposition do not obscure the view: the *Science Allemande* is perfectly reconcilable with his earlier work, and valuable because it offers an overall perspective on that work and on the rôle played in it by Pascal's ideas. In particular it helps to clarify the rôle assigned to *common sense,* the aspect of Duhem's thinking that has caused his philosophical readers most trouble.

It will be recalled that in the passage that closes Part i of the *Théorie Physique,*[8] Duhem insisted that there were no logical grounds available to resist incoherent model-making in physics:

> Indeed we sense that if the true relations of things, beyond the reach of the methods the physicist uses, are in some way reflected in our physical theories, this reflection will lack neither order nor unity . . . He who would see in this no more than a snare and a delusion could not be reduced to silence by the principle of contradiction; but he would be excommunicated by *common sense.* In this circumstance, as always, science would be impotent to establish the legitimacy of the very principles that set out its methods and direct its researches, were it not to appeal to *common sense* . . . Hence, reason has no logical argument to stop a physical theory trying to break the chains of logical rigour; but 'nature supports reason when impotent, and prevents it wandering that far'.

But in discussing the choice of hypotheses at the end of part ii of the *Théorie Physique,* Duhem insisted on the absurdity of using, in the manner of Euler,[9] *common sense* to derive the hypotheses of physics, while, in the previous chapter,[10] he made *bon sens* the judge of whether hypotheses should be abandoned. The consequent temptation to treat this aspect of Duhem's thought as irremediably inconsistent has proved irrestible to many of his readers, but it should be resisted. In the passage just quoted *common sense* is, to use a valuable distinction introduced by modern logicians, a so-called meta-principle, a principle not used in the science itself, but in discussions about that science. Duhem is completely at liberty to insist on the importance of common sense as a meta-principle while at the same time denying it any rôle in the science itself, at liberty to appeal to common sense to support his demand for consistency, while forbidding its use as a source of, or criterion for accepting, scientific hypotheses. Similarly, as will appear below, what is said about *bon sens* and its rôle in physics is perfectly consistent with what is said about *common sense* because the two are carefully distinguished, and indeed further sub-distinguished, being given different sciences to rule. At least two of these are visible in the *Théorie Physique:* that *common sense,* more properly *connaissance commune,* common knowledge, denied any right to judge the hypotheses of physics, and that other *common sense*, a sense of fitness of what is 'sensible' and what is not, whose right to require logical coherence and the rejection of refuted hypotheses is on the contrary affirmed.

That distinction, however, does not exhaust the possibilities. The *Science Allemande* offers a multifacetted classification of the forms of common sense, giving each its own different rôle to play, and these different forms underpin a Duhemian classification of the sciences that in the *Théorie Physique* remains implicit. In the first place there is common knowledge, *connaissance commune,* stated to be the source of the first principles of both pure mathematics and metaphysics:[11]

> As for the axioms, where do they come from? They are taken . . . from common knowledge; that is, every person of sound mind takes it that he is sure of their truth before studying the science whose foundations they are.

and of them we have intuitive certainty, intuitive certainty, for example, of the principles of the geometry of Euclid, since, using an argument that does not seem to be original to Duhem, we all know that objects of the same shape can differ in size, a piece of common knowledge incompatible with every geometry that denies the so-called parallel axiom of Euclid. (Through any point in a plane there can always be drawn one and only one straight line parallel to a given straight line. It is the rejection of this axiom that makes a geometry non-Euclidean.) This position of Duhem's may well be thought wrong-headed: it certainly commits him in advance to rejecting Einstein's General Relativity, dependent as that is on the use of a four-dimensional non-Euclidean geometry. But in so regarding the geometry of Euclid as the intuitively certain true theory of physical space and other 'geometries' as no more than algebraic exercises, Duhem knew what he was about: he was in correspondence with the major figures, such as Eugenio Beltrami, Tullio Levi-Civitā and Felix Klein, who were developing the non-Euclidean geometry of the pioneers Lobachevsky, Bolyai, and Riemann, and he had more than adequate mathematical competence to understand their achievements.

In the second place, varieties of *common sense* have rôles to play in the experimental sciences. There is, for example, "a certain sense of rightness that is one of the forms of *bon sens*";[12] and which guarantees the validity of "the yet imprecise reasonings by which consequences susceptible of experimental testing are derived from a preconceived idea",[13] for such reasonings cannot generally be conducted *more geometrico,* in the geometric manner, for:

> the ideas on which it bears are no longer abstract and ultra-simple concepts like the first objects of the mathematical sciences, notions constructed from these concepts by a well known type of definition: richer in content but less precise and well-analysed, these are the ideas that arise more directly from our observations. To reason accurately on such notions, the rules of the syllogism will no longer be adequate.

In the experimental sciences, *bon sens* has also to play the rôle mentioned above of deciding when hypotheses have been refuted or not. This time, Duhem illustrates his point by envisaging Pasteur (with whose work he was in close contact as a student at the École Normale, and in whose laboratory he might well have made his career instead of continuing in mathematical physics) experimenting on rabbits with an inocculation that should kill them all, though of course some escape death and some of the controls die anyway for a variety of reasons. Such 'failures' obviously pose a problem for any theory of experimental method. Duhem comments:[14]

> Who will decide whether or not these failures are such that the supposition has to be renounced? *Bon sens.* But that decision is all of a kind with the judgement in a trial in which each of two parties is faced with some facts tending to convict him, as well as others tending to clear him: good sense will only bring his judgement after mature consideration of both the for and the against.

and thus *bon sens* is identified as a form of mature judgement.

Yet another rôle for *bon sens* in the experimental sciences is to devise the hypothesis that is to replace the one that has to be abandoned. For that,"in which Pasteur excelled", you have to:[15]

> listen to what each of the observations that condemned the first idea suggests, interpret each of the failures that destroyed it, bring together all these lessons in the construction of the new thought to be thereafter subjected to the measure of reality. A delicate task indeed, in which no precise rule guides the mind . . . to perform it well, *bon sens* must excel itself, and become what Pascal called the intuitive mind. [*esprit de finesse*]

As I have suggested above, this final identification of *finesse* as a superior form of *bon sens* has the effect of putting *bon sens* in the same general category as *finesse*. Duhem was thus able to harmonize the ideas of his Pascal texts into a general theory that a balance between *finesse/bon sens* and *géométrie* are essential to sound work in the sciences, the lack of either leading to incompleteness and inadequacy:[16]

> Just as there are two kinds of minds . . . and each of these contributes . . . to the construction of science, so there are two kinds of orders, the geometric and the natural: introduced in the right place each of these is

> the source of light; but to be content to assign a natural order to the things that belong to the geometric mind is to fall at once into error; and if we asked the geometric mind to illuminate what belongs to the intuitive mind we would remain in deep darkness.

It will follow that the proper balance depends on the subject treated. In the preceding lecture, more concerned with showing the limits of *géométrie* than those of *finesse,* Duhem had denied any rôle at all to *géométrie* in history. There could be no such things as a historical method because there was no deduction in history.[17] It all depends on what you mean by 'method': Duhem took an uncharacteristically crude etymological route to identifying it with deduction, thereby shutting himself off from the real advances in historical studies taking place in his time, in favour of easy destructive criticism of the possible excesses of some historians. But in the final lecture though, the interest is in Duhem's conception of a 'natural order'. A 'geometric order' has to consist in following the deductive sequence of the theorems whereas: "To follow the natural order is to bring together those truths relating to things that are analogous by nature, and to separate judgements relating to different things".[18] The need to do this occurs often and in surprising places:

> Even in geometry it is sometimes necessary to take the natural order into account: indeed, with no loss of rigour, . . . it happens that we can conceive of several different arrangements of the same collection of theorems; in this case, the intuitive mind will indicate . . . which is . . . the most natural and hence the best.

thus making clear that the natural order is judged by *finesse* alone, and is not subject to proof.

If the natural order, and hence *finesse,* has a rôle to play in mathematics, it has *a fortiori* an even greater rôle in physics, and neglect of it gives Duhem grounds for criticizing the work of other physicists, such as his correspondent Waldemar Voigt. As Duhem presents the matter in an extended footnote, Voigt had subordinated the entire structure of his *Lehrbuch der Kristallphysik* to considerations of crystal symmetry deriving from Pierre Curie with, as Duhem saw it, monstrous results:[19]

> The effect of this purely mathematical order is to place in chapters very far apart certain phenomena constantly associated with each other in the thought of physicists. Thus it is with dielectric polarization and magnetization, placed far from each other by this treatise; whereas from Aepinus and Coulomb, the analysis of each of these two properties has never stopped repeating that of the other, and every advance in the knowledge of the one has led immediately to an advance in that of the other.

After that it is inevitable that Duhem will criticize the sterility of the pure mathematics that neglects physics,[20] rejecting as purely geometrical Linnaeus's classification of flowering plants on the basis of the number of stamens alone, and offering corresponding commendation to the natural classification of Jussieu.[21] In sum, and this may throw some light on Duhem's much-misunderstood doctrine of natural classification in physics:[22]

> to obtain a natural classification, it is not enough . . . to make an arbitrary choice of a character lending itself to the language of Arithmetic and carry out a simple enumeration: it is necessary to take all the characters and *weigh* them, so as to find out which exerts the most *influence,* and which the least, on 'the balance of affinities' . . . It cannot be said more clearly that the establishment of a natural classification is beyond the powers of the geometric mind and that the intuitive mind alone can make the attempt.

Thus while our "idea of truth, invincible to any Pyrrhonism",[23] and demanding a natural classification from our theories, is a manifestation of *bon sens* proper, controlled by our sense of fitness, or even common sense itself, carrying the project through is the affair of *finesse.* Whether Duhem here is right or fair in his judgements on Linnaeus and Jussieu is not my concern: this is rather to establish just what Duhem meant by his attitudes to them and what conclusions he drew. What emerges is that Duhem's views on physics are of a piece with his views on other sciences, that criticisms of his physical views are likely to be paralleled by corresponding criticisms of his attitudes to zoology and botany, that the account Duhem gives of physics is a particular case of a quite systematic epistemology, for that, eccentric as it may seem, not amateur psychology and sociology, is what the *géométrie / finesse* distinction is about.

Of the various forms of *bon sens* discussed above, only common knowledge, *connaissance commune*, can really be said in Duhem's eyes to bear the weight of the intuitive certainty ascribed to that form of knowledge and used to build a science, and he specifically restricts it to what he calls "the sciences of reasoning".[24]

> In algebra, in geometry, as also in Metaphysics when it is soundly constructed, the axioms are of extreme simplicity: let our attention be fixed for a moment on some one of them, and all at once, its sense will be perfectly evident to us and its certainty fully assured . . . It is altogether different in the experimental sciences: in these sciences the principles are no longer called axioms, but hypotheses or suppositions . . . Merely attending to the enunciation of a hypothesis or law gives us no right to hold it true: we can only acquiesce in it after the complicated and prolonged task of experimental testing.

In other words, only the sense of accuracy that guarantees our deductions from our hypotheses comes anywhere near the certainty of *connaissance commune*. We are in the domain of judgement, and while judgements may be right or wrong, and thus criticizable, there can be no certainty attached to them, a situation in marked contrast to that in pure mathematics, where, as already remarked, common knowledge gave Duhem his license to reject non-Euclidean geometries as true geometries of space:[25]

> The truth of geometry does not lie merely in the absolute independence of the axioms with respect to each other, and in the impeccable rigour with which the theorems are deduced from the axioms: it lies also, and above all, in the harmony between the propositions forming this chain and the knowledge given by our reason concerning space and the figures that can be constructed in it through that long experiment called common sense.

Ignoring this principle would, in the eyes of Duhem, lead to a situation in which "reasoning had banished reason".

But, though they could not qualify as sources of certainty, the weaker forms of *bon sens* still carried some weight for Duhem, and became the basis of his lifelong crusade against Maxwellian electromagnetic theory and its derivatives.[26] According to Duhem, Maxwell had achieved his electromagnetic theory at the price of numerous illogicalities and violations of the best established laws of electrostatics and electrodynamics. German physicists such as Hertz, instead of taking Helmholtz's alternative, had given Maxwell's equations axiomatic status, and created a theory which, for technical reasons to do with its mathematical formulation that do not concern me here, was in Duhem's eyes inconsistent with the existence of magnets. Physicists had dealt with the problems thus created by ever bolder new theories that continued to fly in the face of common sense and *common knowledge,* and turned physical theories into systems of arbitrary algebraic manipulations, of which Einstein's theory was in Duhem's eyes the supreme and most absurd example.

2. The Apostle of Common Sense

At this date, Duhem's argument will hardly convince many readers, but it has the virtue of reminding us that there was nothing inevitable about the development of modern physics, and that alternatives were and are perfectly possible. Whatever criticisms there may be of the conclusions he drew from this principle, Duhem must surely be right in his claim that truth in physics is about more than algebraic consistency. His argument also shows that to the end he was faithful to the resolve, reported by Picard from a letter to a childhood friend:[27]

> I have believed it my duty as a scientist, as well as my duty as a Christian, to make myself the unceasing apostle of common sense, the only foundation of every scientific, philosophical, and religious certainty. My book on physical theory had no other object than setting out that thesis.

The above point, and its application to the science and faith question, have already played an important rôle in earlier chapters. Considering the supposed (presumably by positivists) meaninglessness of philosophy and religion, Duhem had concluded:

> After much reflection on these difficulties, I realized that the same could be said of all the sciences, and of those regarded as the most rigorous, Physics and Mechanics, even Geometry. The foundations of each of these edifices are formed of notions which are claimed to be understood, although there is no demonstration of them. These notions, these principles, are formed by good sense. Without this base of good sense, in no way scientific, no science could hold up; all of its soundness comes from this.

Take away Duhem's *bon sens,* and we are left with the modern notion of sciences as arbitrary systems of hypotheses with no certainty whatever, with relativism, or with the deductive gap filled precariously by probabilistic confirmation theory. It is needless to remark that this move in no way imperils Duhem's claims about the relative standing of physics, philosophy, and theology. In that sense only is Duhem's methodology "the philosophy of science of a believer", but, as he insisted, the logical point does not depend on his Catholic beliefs.

One other aspect of the methodology of *bon sens* is its Aristotelian flavour, though it was drawn, not from Aristotle, but from Pascal. As expounded by G. E. L. Owen,[28] Aristotle's procedure was to start with the phenomena, in which he included common sense and common observation as well as the carefully tested phenomena of modern science, and then by a critical process to extract the principles of his sciences. Indeed Duhem himself gives a similar account of Aristotle in discussing the analogy between his physics and Aristotle's in the conclusion of *Le Mixte*.[29] Underlying both accounts is the infinite regress problem discussed above, which, as the discoverer of the formal logic on which it depends, Aristotle was the first to articulate. Duhem could have found in Aristotle both the problem and an approach to its solution somewhat like his, and for some readers it has been fatally easy to conclude from external evidence and a superficial knowledge of ecclesiastical history that that was what he did. But people do not always derive their acquaintance with problems and solutions from the source that seems obvious to outsiders: everything we know about Duhem, the circles in which he moved and the culture of his time, points away from Aristotle and towards Pascal, and further evidence of this is presented below. Analogies of the sort discussed in *Le Mixte* are little more than superficial: as Duhem himself pointed out, he was constructing a mathematical physics, which Aristotle emphatically was not. They express little more than what was common between Aristotle and Pascal: a concern with the consequences for knowledge of having a deductive logic. This, however, is not the only area where superficial resemblances lead readers to see in Duhem a Scholastic orientation where none exists. His famous thesis of the autonomy of physics is another, and to this I now turn.

3. Scholasticism, Autonomy, and Pascal

As has already been remarked, Duhem's doctrine of the autonomy of physics has traditionally served as one of the principal grounds for regarding him as a neo-Scholastic, despite persistent neo-Scholastic protests from such as Vicaire and Maritain. This interpretation derives its initial plausibility from the scholastic tradition of talking about the sciences as methodologically distinct separate entities, but that initial plausibility is weakened by closer examination and by the availability of an alternative source of such ideas in Pascal's theory of orders, to be explained below. Nowhere to my knowledge does Duhem use it explicitly, but he does use the language of that theory and examples of it have been cited above. Though by no means identical with Duhem's autonomy doctrine, it is a likely factor in the formation of the latter, to which it offers an instructive parallel. I present below its essential features before coming to Duhem's doctrine.

In a classic fragment, 'Against the objection that Scripture has no order',[30] Pascal asserted that "The heart has its order, the mind its—by principle and demonstration. The heart has a different one. You do not prove you should be loved by setting out in order the causes of love: that would be ridiculous". The thought here expressed that Scripture (and love) are subject to norms other than those of logical deduction is basic to Pascal's apologetic enterprise, but, as another fragment makes clear, this is by no means its only application:[31]

> The infinite distance of bodies from minds symbolizes the infinitely more infinite distance of minds from charity, for that is supernatural.
>
> All the glory of greatness has no lustre for those engaged in the researches of the mind.
>
> The greatness of the people of the mind is invisible to kings, the rich, commanders, and all those who are great in the flesh.
>
> The greatness of wisdom, which is nothing without God, is invisible to the fleshly and to the people of the mind. These are three generically different orders.

Pascal's distinction then, refers both to things and to how and by whom they are known: it is both ontological and epistemological. Each of his orders is subject to its own norms and is to be investigated by different methods, or by different kinds of mind, a conclusion also implicit in the opening of a classic Pascal fragment, the text Duhem used to set up the argument of Part i, Chapter iv of the *Théorie Physique:*[32]

> Various kinds of sound minds: the one kind in a particular order of things and not in the others, where it wanders off course.
>
> The one is good at drawing the consequences from few principles, that being one kind of soundness of mind.
>
> The other is good at drawing the consequences in which there are many principles. For example, the one kind understands well the effects of water, in which there are few principles . . . and for that reason these would not be great geometers, because geometry involves a great number of principles . . .

The details of Pascal's argument here are not clear—we are not used to thinking of geometry as a subject in which there are many principles and hydrostatics as a subject in which there are few—but some things are clear enough for my purposes: the number of principles in a subject is a consequence of its nature, not of human ingenuity; and Pascal's tendency to divide the world and our knowledge of it into different orders, despite its usual apologetic reference, is by no means tied to the theological area: on the contrary it is fundamental to his intellectual outlook. In any case, it was a constant feature of his thought from the beginning of his career, and Krailsheimer[33] sees it as originating in Pascal's mathematical work. It is clearly stated at the beginning of the *Art de Persuader,*[34] and, though without its specialized terminology, it is already visible in the *Préface sur le Traité du Vide*[35] (written about 1651) in which Pascal complains that:

> . . . in matters in which it ought to have the least weight the respect

> accorded to Antiquity has reached such a point today . . . that we cannot any longer put forward novelties without peril, and the text of an author is enough to defeat the strongest arguments . . .

For:

> In those matters in which the object is simply to know what authors have written, as in history, in geography, in jurisprudence, in languages, above all in theology, and finally in all those whose principle is either the bare fact or divine or human institution it is necessary to turn to books, since everything that can be known is contained in them. Hence it is evident that we can have complete knowledge of them, and that it is impossible to add anything to them . . .

Whereas:

> it is not the same with things subject to the senses or reasoning: authority is useless here; reason alone is in place. They have their *separate* rights: once the one had the advantage; now the other reigns in its turn. but because the subjects of the latter kind are proportioned to the scope of the mind, it finds total liberty in extending itself in them: its inexhaustible fertility yields continually, and all its inventions together can be without limit or interruption . . . [my emphasis]

Modern readers will perhaps be more sympathetic to Pascal's vision of limitless scientific discovery, contrasting marvellously with Bacon's utopian vision of a finite task to be completed in a generation, than with Pascal's static authoritarian vision of history, law, and theology. But irrespective of the grounds of Pascal's *separation* of these different orders, or of his view of its nature, the idea is by no means dead, as witness the continuing debates about whether history and anthropology can be scientific, or about whether methods derived from other sciences can be appropriately applied to them. Duhem's *separation* of physics from metaphysics can easily be seen in the context of this kind of debate.

4. Pascalian Orders and Duhemian Autonomy

According to Pascal, the properties of water were of a different *order* from those of mathematical systems, requiring a different kind of mind to understand them, just as for him the order of charity was quite distinct and separate from that of mind, and that of material things. It does not necessarily follow that Pascal would have understood or approved of Duhem's particular separations of physics from the other sciences, of physics from metaphysics and physics from common sense, but it seems possible: Eastwood seems to think so anyway:[36]

> Both for Pascal and Duhem the indemonstrable character of first principles . . . deprives it [scientific knowledge], not only of metaphysical support, but, conversely, of any claim to treat human destiny and the inner meaning of things.

In this she relies on Gustave Lanson's view that in his scientific work Pascal:[37]

> proceeds by a rigorous method, chosen according to the object, distinguishing carefully what is appropriate to mathematics from what is appropriate to physics, very careful here to ask for no more than the evidence of reason, and there to use no more than the observation of the senses, rejecting all philosophical systems and metaphysical ideas, only recognizing determinate problems treatable according to their nature by particular methods, very different even here from Descartes, who knows only one universal method.

However that may be, whether or not Pascal makes precisely the same distinctions as Duhem, a man steeped in his writings is likely to find himself making distinctions *of the same kind* as those to be found in his favourite texts. In this he will need no encouragement from the neo-Scholasticism of his own day. Moreover, the very form of Duhem's doctrine of the *autonomy* of physics, of its *separation* from metaphysics, decisively separates it from Scholastic theories, aligning it with Pascal instead.

His core objection, set out in the first chapter of the *Théorie Physique*[38] to the search for explanation in physics is that it subordinates physics to metaphysics, with the result that the standing of a physical theory depends, not on the facts, but on the metaphysical system favoured by physicists, just as in fact happened in the seventeenth-century argument about occult qualities. This arose from the reaction against the Scholastic method of supposing that behind the properties of observable things there lay *occult* qualities which could be grasped by essential definition. Following Descartes and Gassendi, the 'new philosophers' of the time proposed to admit only those 'primary qualities' that could be handled by a mathematized theory of the motions of bodies. There was, however, some dispute about just what these primary qualities were, and so it came about that, for example, Cartesians and others rejected the attraction hypothesis fundamental to Newton's *Principia* because their metaphysical systems could make no sense of such a hypothesis, whatever its compatibility or otherwise with the facts. So it was that in that period each rival school of natural philosophy accused the others of reviving the occult qualities of scholasticism. Duhem makes this point briefly in his opening chapter, and then enlarges on it when discussing the choice of primary qualities in Part ii of his work.[39] There he poses a characteristic rhetorical question:

> The physicists who sought to construct explanatory theories took, from the philosophical precepts they submitted themselves to, touchstones and reagents capable of detecting whether the analysis of a property had penetrated to its elements . . .
>
> For us, who make no claim to explain the properties of bodies, but only to give a condensed algebraic representation of them, who do not appeal to any metaphysical principle in constructing our theories, but understand how to make physics an autonomous doctrine, where would we find a criterion allowing us to declare this quality truly simple and irreducible, that other quality complex and in need of a more penetrating dissection?

As Duhem emphasized in 'Physique de Croyant',[40] the doctrine of natural classification makes no difference to this situation: as explained in my previous chapter, it permits between physics and metaphysics no more than analogies, analogies subject to *finesse* and not to *géométrie,* to intuitive and not to logical considerations.

Armand Lowinger[41] considered the autonomy doctrine the core of Duhem's position. Roberto Maiocchi[42] has shown on the contrary that it was merely positivist conventional wisdom when Duhem was writing: Duhem was merely doing what was expected of a late nineteenth-century positivist. If not 'neo-Scholasticism', 'positivism' would have had to be the label for his theories. Duhem, however, put up equally sturdy barriers between physical theory and experience: for him the comparison between physical theory and experiment is ultimately dependent on intuitive judgement, not on logical deduction, and the point is rubbed in with strictures on attempts such as those of Euler to deduce the hypotheses of physical theory from axioms supplied by common knowledge. Not only did he erect thereby a barrier between physics and common sense, but he also reinforced that between physics and metaphysics, since, as explained above, Duhem held that the principles of metaphysics *were* derived from common knowledge. But surprising as it may seem at first blush, none of this is incompatible with those other Duhemian principles, that common sense is the regulative principle of all knowledge, and that physical theory should not violate common sense: judging such violations is not a matter of deduction.

5. Scholastic Contrasts to Duhem's Autonomy Doctrine

Much of this is a commonplace of Duhem commentary, but less often understood is how strongly it contrasts with a standard account of neo-Scholastic inspiration. By way of comparison I quote from a particularly clear *New Catholic Encyclopedia* article by E. D. Simmons:[43]

> St. Thomas shows how sciences, though different, can be interrelated. For example, a given science can be different from another and yet included under it or subalternated to it. This is the case when one (higher and subalternating) science supplies the reason for the fact established in another (lower and subalternated) science. In natural philosophy, for example, reasons are given for things that are seen as facts in medicine, and in arithmetic, reasons are given for things that are seen as facts in music. St. Thomas also recognizes the existence of sciences that are noetically mixed, that is sciences that are formally mathematical in their mode of demonstrating, but the subjects investigated in them are physical.

Simmons, it will be noted, here uses words like 'physical' and 'mathematical' in their peripatetic senses, corresponding respectively, as Duhem noted in *To Save the Phenomena,*[44] more nearly to 'metaphysical' and 'physical' in modern usage. He also ascribes to Thomas a scheme patently Aristotelian in inspiration, and as such a good illustration of the way in which neo-Scholasticism was taken to imply Aristotelian methodology and natural philosophy, with all the difficulties these implied for the practitioners of a modern science. It actually, however, violates one important norm of Aristotle's method: since for Aristotle demonstrations had to stay within the boundaries of the natural kinds, reached by essential definition, that were the objects of his sciences. Nevertheless it was one way, which seems to have originated in the ancient world, of adapting Aristotle's methodological principles to allow for mathematical sciences, and as such it was influential, and Simmons's article gives us a clear statement of it. The same set of ideas lies behind Jacques Maritain's *Distinguer pour Unir, les degrés du savoir* (Distinguish to Unite, the degrees of knowledge), which aims to unite all the sciences it distinguishes into an over-arching hierarchy of knowledge, in which:[45]

> Every superior discipline is regulative with respect to inferiors. Since Metaphysics considers the supreme *raisons d'être,* it will be the regulative science *par excellence, scientia rectrix.* But mathematics too is a deductive science, a science of the causes of things. It too will tend to regulate the lower parts of knowledge, or even to *transgress* on to the territory of Metaphysics itself.

The content of Maritain's regret appears from my next quotation:[46]

> the great discovery of modern times . . . is the . . . possibility of a universal science of nature informed not by philosophy but by mathematics: . . . a *physico-mathematical* science. This prodigious invention—without . . . changing anything . . . in the essential order of the things of the mind— . . . has . . . been the occasion . . . of the terrible misunderstanding that for three centuries has set modern science and the perennial philosophy at odds with each other. It has caused great metaphysical errors. . . . Of itself . . . it is an admirable discovery, and in any case it is easy to assign it its place in the system of the sciences. It is a middle science . . . *materially physical* and *formally mathematical.*

Even if we disregard Duhem's repeatedly stated intention to stay "constantly within the limits of physical science",[47] such language contrasts strongly with Duhem's usual tone and style. If Maritain represents the orthodox neo-Scholastic Thomist position, Duhem's views must by contrast inevitably be seen as subversive, and it is hardly surprising that the *Revue de Philosophie,* despite Duhem's association with it, should have carried extended multi-part articles claiming to justify, specifically against Duhem, the scientific value of metaphysics, examples being 'Symbolisme et Liberté dans les Sciences' by R. Marchal S.J. and M. Gossard's 'Sur les Frontières de la Métaphysique et des sciences'. Maritain's criticisms of Duhem will be discussed in my final chapter.

In contrast, Pascal's theory of the orders does not carry any necessary implication of subordination or unification. While his *géométrie* and *finesse* or *sentiment* have sometimes to cooperate—as when the latter has to provide the principles for the deductive sciences—there is no attempt to create a unified system of all knowledge. Another quotation from the already cited *Préface sur le Traité du Vide* will emphasize the point:[48]

> The clarification of this difference ought to lead us to complain of the blindness of those who introduce bare authority in physical matters, instead of reasoning or experiments, and give us a horror of the malice of others, who employ reasoning in theology instead of the authority of Scripture and the Fathers.

But, it will be said, Pascal too had his system of the sciences. To accept this point, however, is at once to recognize what a different sort of system that was. Maritain's system subordinates the *content* of inferior sciences to that of superior sciences, since, it is his claim, the latter establish (deductively) the reasons for the facts established as facts in the former. Pascal's system in no way links the *content* of the sciences of the different orders to each other in this way. I have already referred to the distinction established by modern logicians between object-level talk about objects as such, and second-level talk about theories of such things: Maritain's system operates at the first level, Pascal's and Duhem's at the second. Moreover, Pascal's system, and the system Duhem built on it, obviously and persistently face the other way from Maritain's: *finesse* and *géométrie,* intuitive judgement and deductive reason, which provide the common linking themes, serve not to link the different sciences together—the sciences of reason, experimental sciences, and historical sciences—but to illuminate their differences. Pascal and Duhem distinguish to separate, not to unite. Whatever may be the common source in Aristotle or earlier of all such distinctions between the sciences, there really is not enough common ground to link Duhem to any post-mediaeval scholasticism. Or if Duhem's thought is Scholastic, it will emerge later, his Scholasticism will be that of William of Ockham, not of Aquinas.

VI

The Shape of a Pascalian Methodology

1. Composition and Origins

In my previous chapter I argued that lifelong meditation on certain texts of Pascal shaped many of the most important and difficult features of Duhem's thought. Particular examples are the emphasis on intuitive elements in our knowledge and the doctrine of the autonomy of physics, the former most fully developed at the end of his life with the explicit epistemology and methodology of *bon sens* in *Science Allemande,* but both prominent in the *Théorie Physique* of 1906. The latter work, however, is the main centre of scholarly interest in Duhem, even where the chief concern is historical rather than philosophical. If Pascal's work really is the key to interpreting Duhem, it has to provide a coherent interpretation of the entire work, not just parts of it.

In this chapter, I present an analysis of Duhem's work as built up from a subtle dialectic of *géométrie* and *finesse,* of logical and intuitive factors, which can easily mislead the unwary. In a manner familiar in the rather cruder 'straw man' technique of philosophical argument, Duhem erects positions in order to pull them down again, erect others, and demolish these in their turn. He goes back on his tracks and, in a manner recalling that of the *Pensées,* discusses the same issues from differing points of view. However, all this is complicated, as it often is with Duhem, by the publication history of the work, and something has to be said about this before I discuss its structure.

With its 415 pages in the second edition of 1914, *La Théorie Physique, son objet et sa structure* is a substantial book, much longer than would be expected of present-day books in this area. The sub-title gives a general view of the structure of the book itself: it has two parts devoted respectively to the aim of physical theory and its structure. Into the former go questions of explanation and classification, coherence and the search for it; into the latter go questions about the rôle of mathematics, the interpretation of experiment and the source of hypotheses. The second edition also reprints a couple of articles[1] previously published elsewhere, and both related to the work of Duhem's younger contemporary the philosopher Abel Rey. Like many of Duhem's other works, this one was first published in serial parts in a journal, in this case the *Revue de Philosophie,* starting in April 1904 and continuing thereafter at approximately monthly intervals until it was completed in the early summer of 1905, giving 13 instalments in all. In view of what is known to have happened in the case of the *Origines de la Statique* (discussed below in Chapter VIII—Duhem composed it as he went along), it cannot safely be assumed that the manuscript was complete before publication began. That it was not would be consistent with the appearance of the book in 1906, some six months after serial publication was complete—otherwise it could have been set up in print ready for when the *Revue* had finished with it. An interval of over a year between commencement and completion would leave room for second thoughts, second thoughts, for example, that could even have affected such crucial matters as the definition of a physical theory in *Théorie Physique.* Early on Duhem states that it is a system of mathematical propositions with the function of representing, not explaining, a set of experimental laws. He then offers a list of the operations needed to set such a thing up:[2]

1. The definition and measure of physical quantities.
2. The choice of hypotheses.
3. The mathematical development of the theory.
4. The comparison of the theory with experience.

The list is logical, not to say 'geometrical', and Duhem might have been expected to follow this order in Part ii of his book. But he does not. He begins with 1, discoursing on quantity and quality and the selection of primary (= basic) qualities as if he really was about to set up in accordance with this scheme an energetics or general thermodynamics of the kind referred to in my Chapter I. But then he skips 2, goes straight to 3 and 4, leaving 2 till last. His reason for skipping 2 is interesting:[3]

> . . . before sketching out the plan of the foundations that will carry the edifice, before choosing the materials it will be built with, it is indispensable to know what building it will be, to become acquainted with the pressures it will exert on its bases.

To my mind, that argument could just as well have justified skipping 1 as well. Moreover, his treatment of 3 is not clearly distinguishable from that of 4, and jointly the two dominate the latter part of the work as they have dominated Duhem commentary ever since. The reader is entitled to wonder just what is going on. But there is a deep reason for this shift: the list of tasks for the constructor of a physical theory is a consequence of a definition Duhem himself is going to supersede, or at least supplement, when he introduces his motion of natural classification. As it stands before modification, it corresponds to a notion of physical theory as the theoretician's arbitrary invention. But after the introduction of natural classification physical theory ceases to be an arbitrary creation: it becomes instead the product of a historical development, hardly to be understood apart from that history. Moreover it is now a step on the road towards the hoped-for goal of a natural classification in the future. Modifying the notion of a physical theory in this way was bound to have consequences for the natural order of exposition, consequences that really needed noting in advance to avoid misunderstanding, but Duhem did not so note them, nor does he seem to have realized the intimate relationship between the definitions of the quantities treated by a physical theory and the hypotheses.

But the complexities of the composition of the work go far beyond those connected with the circumstances of its serial publication in 1904–05. Its conception was not the work of those years, but of ten years earlier: as already hinted above, it originated in a series of articles published in the *Revue des Questions Scientifiques* in 1892–94,[4] articles soon so famous that Duhem's ideas dominated contemporary debates on scientific method. Being derived from these articles, much of the presentation and argument, if not actual wording, of the *Théorie Physique* was over ten years old, bearing all the marks of the debates of the earlier period, with only partial updating to cover later developments.

An example of this is the central claim of Part ii of the work: that physical experiments cannot condemn isolated hypotheses, but only theoretical systems. Many readers must have been puzzled by the example used to support Duhem's argument here, Otto Wiener's experiment on the direction of the plane of polarization in the vibratory or wave theory of light,[5] puzzled because, though Duhem writes about it as if it was a matter of current interest, at the time he was writing this was surely a dead theory, long superseded by the electromagnetic theory, and as such uninteresting to active experimenters and theorizers in physics: even for modern contextualist historians of science, Wiener's experiment seems of esoteric interest only. However, Gaston Milhaud provides an explanation of this anomaly: this experiment was the subject of an enthusiastic report to the Académie des Sciences in February 1891 and the comments of Poincaré referred to by Duhem were made in the ensuing debate.[6] Cited against him by Vicaire in 1893, it is unsurprising that Duhem should in his turn have chosen to discuss it in a paper published in July 1894 and written earlier in that year at the latest, when he had just transferred to Rennes from Lille and was negotiating a second transfer.

Though possibly reprehensible, it is even understandable that when he came to reuse the material in 1904–05, Duhem should have omitted to refer to his own final view of the significance of the experiment, a final view with many implications for the interpretation of his philosophy of science, and to be found in his essay review of the 1903 facsimile reprint (by A. Hermann) of N.M. Ferrer's edition of *The Mathematical Papers of George Green.*[7] The reprint tells its own tale about the interests of French physicists of the period. The vibratory theory of light, associated with the names of Thomas Young and Augustin Fresnel, was essentially a development of the mathematics of elasticity, the focus of Green's work, in which it was supposed that a light-bearing elastic ether vibrated in a plane at right angles to the direction of travel of the light rays. This *transverse* vibration was needed to account for the well-known phenomenon of the polarization of light, a phenomenon since made use of in one system of three-dimensional cinematography, and in spectacles permitting objects to be seen under water from above. The problem was whether the direction of this transverse vibration should be parallel to the plane of polarization or at right angles to it. As Duhem saw it, the elastic (vibratory) theory of light had always suffered from a basic contradiction: some applications required the plane of vibration to be parallel to the plane of polarization, but others required it to be perpendicular. Duhem's attitude, whose implications will be discussed below, is quite unambiguous: the vibratory theory of light has been refuted.[8]

> As early as 1835, F. E. Neumann pointed out this contradiction. This is not the place to examine the method he proposed of resolving it: that would drag us into a complete discussion of the relation to be established

> between the planes of vibration and polarization. Such a discussion, to which O. Wiener's experiment has added an essential element by imposing a new condition on to the variables of the problem, would lead us to reject as inconsistent the elastic theory of light.

The leading ideas of the *Théorie Physique* were thus almost all, in one way or another, given expression in this series of articles, with, as was only to be expected, some differences of emphasis and development—natural classification, for example, is here to be found only in germ. The starting point, the article 'Réflexions' of 1892, apparently represented the opening lecture of his Lille theoretical physics course. As I mentioned above in Chapter III, its tone is empiricist, not to say Machian: assistance to the memory is the object of physical theory; it is for sorting out the confused mass of empirical data; logic places no restrictions on the choice of hypotheses. However, with a confidence on this subject he was never to show again, Duhem asserted:[9]

> In fact, it is absolutely certain that this choice is not random: there are general methods followed in choosing the fundamental hypotheses of most theories; to classify these methods is at the same time to classify the theories.
>
> The ideal and perfect method would consist in choosing no hypotheses other than the symbolic translation into mathematical language of some of the group of experimental laws it is desired to represent.

Of course the ideal and perfect method is impossible: Vicaire was to attack him mercilessly for even suggesting it, but Duhem knows it anyway: "Let us say at once," he writes, "physics offers us several theories more or less closely approaching this ideal, but none fully realizing it",[10] and goes on to refer to the examples of Newton and Ampère he was to analyse so mercilessly in the *Théorie Physique,* but at this point he retains it as an ideal.

It is not difficult to see the source of such language in the passage in the *Esprit Géométrique,*[11] discussed in Chapter IV, where Pascal presents his version of the infinite regress argument, a passage which Duhem will refer to again in the *Théorie Physique.* There, as I have explained, Pascal introduces his discussion of geometrical method, via the *admittedly impossible* method of complete deduction of every proposition advanced, and complete definition of every term used. Nevertheless, by introducing such language into this article, Duhem gave hostages to fortune: The extended criticism of Vicaire in 1893, and of others in other Catholic periodicals, forced him to revise and expand his position. Vicaire's suggestion[12] that a random collection of mnemonics rather than a coherent physical theory might be more useful as an assistance to the memory seems in particular to have drawn blood, for Duhem's response to it was the new idea of natural classification. The fruits of the revision and expansion were the articles 'École Anglaise' and 'Métaphysique' of 1893, and 'Physique Expérimentale' of 1894, material that, though re-arranged and expanded, was almost all to reappear largely verbatim in the *Théorie Physique.* The first two forming Part i, Chapters i–iii and iv respectively, and the third Part ii, Chapters iv, v, and vi. The completely new material concerned the mathematization of qualities, the consequences of mathematical deduction and the choice of hypotheses. Noteworthy are the inceased emphasis on *common sense* and *finesse,* and the development of the theory of natural classification.

As a method of composition, this kind of minimal reworking of older material, word-processing before the fact, is by no means unique to Duhem, and must have had its attractions to a man as busy as he was, and who suffered chronic ill-health. Nevertheless, its dangers are easy to see. The influence of the older material available is liable to disturb the balance of the whole, and no matter what care is taken in re-arranging and rewriting, the result may not be entirely coherent. Thus, for example, between Duhem's analysis of physical experiment and that of the relation of theory to experiment (why separate chapters?), Duhem has a chapter on the notion of physical law (sharply distinguished from "experimental law"). In this chapter, physical laws are considered to be theoretical, not practical or experimental, entities. But then, it would seem, they ought to have been considered along with the properties of mathematical deduction as part of the structure experiment has to test. It would also seem natural to wonder whether the analysis of the nature of experiment ought to have come first, to help put the succeeding argument into perspective.

Problems like this show the damaging effects of corner-cutting in the composition of the work. The result is that the *Théorie Physique* is seriously flawed, and worse, as I shall argue in Chapter X, open to serious misunderstanding: particularly when it first appeared, some of its readers were familiar with the earlier papers and saw it merely as republication of them, missing entirely the significance of the additions and changes of emphasis. This danger would be increased by the availability of some of the additional material elsewhere in works like *Le Mixte,* and *Mécanique.* Nevertheless, flawed though it may be, it is, rightly understood, a flawed masterpiece.

2. A Dialectical Strategy for a Pascalian Argument

According to the account I have given earlier, Duhem maintained that deductive reason or *géométrie* alone could not establish any science, whether rational or experimental, and that any sound work needed the assistance of intuitive factors, such as common sense, *bon sens* and *finesse,* with pride of place to the latter. I now claim that in the *Théorie Physique* Duhem establishes the insufficiency of deductive reason by a dialectical argument in which the successive attempts of deductive reason to impose its empire on physics are defeated, leaving *finesse* or intuitive judgement in charge. This dialectical structure means that points established early in the work do not necessarily stand for Duhem's final position: like the initial hypotheses of a *reductio ad absurdum* proof in geometry, they may only be there for subsequent modification or refutation. In the name of *bon sens,* Duhem is going to reject any position that would leave physical theory completely determined by pure logic, and at the same time, in the name of *common sense* he is going to reject any theory that excludes logical ordering. In arguing for the balance of deductive reason and intuitive judgement in physical theory, the *Théorie Physique* is going to exemplify that doctrine in its own structure.

This dialectical structure has another consequence, that Duhem will keep coming back to the same topics in different or related forms. Thus he will keep returning to the primacy of intuitive factors, to the rôle of deduction, to the simultaneous advantages and disadvantages of combining experimental method with mathematical structure, and this too is Pascalian. For we are under the empire of Pascal's *coeur,* or heart. The heart, as Pascal said in a fragment already used above, has its order, which is not that of the mind or intellect.[13] Just as Pascal had sought to humiliate "reason that is so proud",[14] so Duhem seeks to show the limitations of deductive logic, to show that, as he put it, it "is not the sole guide of our judgements".[15] For, as Pascal put it:[16]

> J[esus]-C[hrist] and St. Paul have the order of charity, not that of the mind, because they wished to humble, not instruct, and St. Augustine is just the same. This order consists principally in digressing on every point that concerns the aim, so as to be constantly showing it.

There are obvious disadvantages in this method of constructing an extended argument—if you want to write for an audience used to a more didactic mode of presentation, or if your book falls into the hands of such an audience and you do not warn readers what you are about. It looks as though Duhem may have lost out on both counts: even in Duhem's time readers conditioned by familiarity with Pascal were in a minority, while both the scholastically oriented and the positivists were used to a didactic presentation, and since then Duhem has fallen into the hands of philosophers with a professional commitment to didacticism, philosophers for whom Duhem's subtle dialectic is liable to seem merely inconsistent. A dialectical presentation of the kind adopted by Duhem may well have limited comprehension to those who shared his Pascalian tastes, and perhaps also to Protestants familiar with Kierkegaard. Furthermore, the presentation may well have served to strengthen the suspicions of many of his readers faced with any talk of factors that escape reasoning, and so helped to deny Duhem the kind of sympathetic attention he needed if his argument was to be appreciated as a whole. A dialectical presentation, however, is an obvious choice for anyone wanting to make use of sceptical arguments, whether to arrive at sceptical conclusions or for other purposes. Readers need to appreciate that it rests not on unquestion*able* principles, but on unquestion*ed* ones, on hypotheses adopted for the purposes of the argument, and that it will proceed via intermediate results to the real conclusions. They have thus to be alert to identify what claims fall into what category. Attention to this point should be enough to avoid gratuitous accusations of inconsistency.

The dialectical structure of Duhem's argument is already visible in Part i. Thus, for example, Chapter i proposes and demolishes the view, ascribed by Duhem to Descartes, but equally applicable to neo-Scholastics, that the aim of physical theory is explanation, and gives two basic reasons, both intimately related to the principle of the autonomy of physics, considered in my previous chapter: the resulting subordination of physics to metaphysics, and the chaos this would cause in the development of physical research, *plus* the powerlessness of metaphysics to *deduce* anything physically useful.[17] The argument is not very thorough or detailed; as remarked above these were the commonplaces of Duhem's time, the agreed conclusions of a century of debate in and out of positivist circles.

History (intuitive judgement) in the form of the argument over occult qualities has already been adduced to illustrate the subordination of physics to metaphysics while demonstrative deduction has already taken its first knock. In the second chapter Duhem offers a replacement for the ideal he has demolished: physical theory as a purely representative mathematical system, with no explanatory pretensions, with experiment the sole criterion of truth. It might be thought that this was Duhem's final position but it is not: as Maiocchi has emphasized,[18] it was a commonplace of nineteenth-century positivism: Duhem is not going to write books to urge such commonplaces, and, as noted above, it will be significantly modified in the sequel. Duhem lists the operations needed to construct such a physical theory, and promises that each of them will get detailed discussion in the remainder of the work, and though he talks of the eminently Machian uses of a physical theory so defined—economy of thought and classification—he then proceeds[19] to introduce his new doctrine of natural classification, a doctrine with, explicitly, no justification in the experimental method, but in Duhem's view essential to the progress of physics. In any case it is essential to his famous attack on incoherent model-making.[20] It begins to appear as if what Duhem has done in Part i is to refute both the main existing views of the object of physical theory, *both* classical realism *and* classical positivist instrumentalism: that at any rate is Maiocchi's plausible thesis.[21] It is worth looking in a little more detail at Duhem's route to this point.

3. Explanation or the Representation of Experimental Laws

In the first chapter the key idea, a very seventeenth-century one, which looks like the product of Duhem's Pascalian perspective, is the need for a criterion of when an explanation explains. Explanation is, says Duhem "to strip reality from the appearances . . . so as to see this reality naked and face to face."[22] This, he says, assumes two things, that there is a reality behind the appearances, and that we know the nature of that reality, both of these being answers to metaphysical questions. Duhem merely asserts:[23]

> . . . these two questions . . . do not pertain to the experimental method; for it knows only the sensible appearances and can discover nothing beyond them. The solution of these questions transcends the observational methods used in Physics: it is the object of Metaphysics.

Duhem's reasons for this claim can be inferred from his analysis of experiment in Part ii, in that if non-explanatory hypotheses cannot be proved by induction or elimination from experiment, it is *a fortiori* unlikely that explanatory ones can, and simply incredible that metaphysical claims could. In the meantime the quoted claim functions as a hypothesis on which the rest of the argument can proceed, but a hypothesis resting on nothing more than a decision to consider what follows from adopting an autonomous experimental science as the primary object.

In Chapter ii Duhem set up his non-explanatory alternative: "a system of mathematical propositions, deduced from a small number of principles, whose aim is to represent as simply, as completely and as faithfully as possible, a collection of experimental laws." However, he now goes back on his tracks and introduces a goal that the experimental method admittedly cannot justify: a natural classification of the phenomena. Natural classification is distinguished from explanation because it is to be sought, not in the *content of the hypotheses,* but in the overall *arrangment* the theory gives to the phenomena, so that there is no inconsistency with Duhem's earlier claims in Chapter i: this had done no more than deny a *deductive* link between physics and metaphysics, a logical link that the new idea does not reintroduce by the back door; it is a substitute for truth and explanation not open to the same criticisms as truth and explanation as originally presented. But, like them, it presupposes things for which experimental method can give no justification. It is based on a conviction to which, no matter how hard he tries, Duhem claims, the physicist cannot refuse his consent, but of which he:[24]

> can never give an account; the method available to him is limited to the data of observation. Hence it cannot prove that the order established among the experimental laws reflects an order transcending experience, and cannot either, *a fortiori,* suspect the nature of the real relations those established by the theory correspond to.

In Pascal's words, it is a reason of the heart "that reason does not know",[25] and the same applies to the rôle of successful prediction in the progress of physics. Duhem's alternative account of physical theory has been rejected and replaced by another, replaced because it had to be: it was an affair of pure deductive reasoning, and had to be brought under intuitive control.

4. Common Sense and Methodological Justification

The justificatory rôle given to common sense at the end of Chapter iv has been discussed in my previous chapter. Two aspects of it are worth noting: Duhem's insistence on its fundamental importance for all knowledge despite his awareness that common sense can be a rather confused business; and that what it establishes is the necessity of deduction. For the first, Duhem can speak for himself:[26]

> At the bottom of the most clearly enunciated and rigorously deduced doctrines we always unearth this confused collection of tendencies,

> aspirations and intuitions. No analysis is penetrating enough to separate them from each other, to decompose them into simpler elements, no language precise enough and supple enough to define and formulate them. Nonetheless the truths this common sense reveals to us are so clear and so certain that we can neither ignore them nor call them into doubt. Moreover, every scientific certainty and clarity is a reflection of their clarity and a prolongation of their certainty.

For the second, it is enough to recall what I have said about this text in my previous chapter, and what it establishes: the only ultimate sanction against those who would violate logic and produce incoherent theories is excommunication by common sense. Common sense, the queen of Duhemian epistemology is required to ground deductive rigour, for the latter is as unable to justify itself as the hypotheses it needs to get to work on. At the same time common sense also demands deductive rigour. Deductive reason, impotent on its own without the aid of common sense and intuitive judgement, is in its turn required by the common sense without which it is incomplete. To justify the certainty and clarity of science, then, we have to appeal to what is admittedly confused, and what is admittedly confused is actually ultimately clear. I hear my readers protest, but perhaps they may recall the ultimate problem behind this: infinite regress. If we think we know anything at all, then it looks as though there is nothing better. Duhem concedes what some intuitionist philosophies have not, that common sense is actually not very clear, that what is clear has to be quarried out of it and subjected to a long process of clarification. What is clear has ultimately to be obtained from something that is not. So, then, do we concede that we have no certainty or probability of any kind? The resistance of philosophers to this conclusion has been the motive force behind much philosophy now and in the past, and behind much of the resistance to the fallibilism of Popper, but one may wonder whether the results of this resistance are any real improvement on what Duhem offers here. In this kind of epistemological morass, readers may perhaps begin to see the wisdom of a non-didactic, dialectical style of presentation such as that used by Pascal and Duhem: it at least offers the possibility of a closer approach to what is ultimately unclear and maybe even inexpressible. The Pascalian strategy of "continual reversal of the for into the against"[27] may have some merits. In Part ii the story is much the same.

5. Intuition and Deductive Rigour in the Structure of Physics

By making common sense the only ultimate guarantor of deductive order and rigour, Duhem has, quite deliberately, subjected physics to a perpetual tension between intuitive and deductive factors, a tension he will enlarge on in his final chapter. As Duhem conceives the issue in Part ii of his work, the need for deductive rigour in physics is met by the adoption of mathematical language, the most apt means of helping the physicist in the extremely difficult task of eliminating the logical fallacies that continually creep into his arguments. This tension is illustrated in his first chapter. On the one hand, as Duhem points out against Euler,[28] we can have a *qualitative physics,* a physics that gives full value to the colours, smells, and textures of our commonsense world, inevitably Aristotelian in form and content. Alternatively, mathematics wins and we have a variant of one of the *mechanical* philosophies, current since the seventeenth century, in which all phenomena are to be reduced to the laws of motion, and based on pure quantity. There is a good reason for this tendency of mathematical theories towards purely mechanical explanation: the core subject-matters of mechanics are the easily mathematizable *quantities* extension, motion, and time, while, on the face of it, qualities are not quantities and so cannot lend themselves to mathematical treatment. But if a method of handling qualities mathematically were available, then there would be a prospect of a mathematical physics that did justice both to common sense and to deductive rigour. Duhem's way of dealing with the problem is to suggest that while qualities as such may not be be handled mathematically, their intensities can be: red is in no way a quantity, but since it can be more or less intense, we can represent the degree of redness of a red object by a mathematical symbol.[29]

The achievement of a mathematical treatment of qualities leaves open the selection of which ones to treat as primary, and fit for basing the theory on, and which to treat as secondary and derived from the primary qualities. For Duhem, this will inevitably be an affair of intuitive judgement. I have already cited from this chapter the passage in which Duhem points out that in the absence of a metaphysical theory, such as that available to physicists who seek explanations, he has no easy guide to the decision as to whether he has too many, he can offer no improvement on Molière's peripatetically-minded *doctus bachelierus* who happily invented dormitive virtues to explain the sleep-inducing powers of opium, or to that learned gentleman's atomistic equivalents. Duhem offers only Lavoisier's definition of a chemical element:[30] that which analysis has not yet been able to break down. Later methodologists, like Popper and Lakatos,[31] have constructed elaborate methodological systems with explicit criteria or methodological conventions purporting to help scientists decide when to reject a proposed new quality or hypothesis as *ad hoc*. Duhem offers no such criteria. Instead, he can only, again foreshadowing his final chapter, offer us a tension between theoretician and experimentalist, the activities of the one leading to a reduction in the number of primary qualities, the activities of the other working in the other direction:[32]

> Chemistry sets out a collection of about a hundred corporeal material things irreducible the one to the other; to each of these material things, physics associates a form capable of a multitude of various qualities.

Each of these sciences attempts to reduce the number of its elements as far as it is able, and yet, as its progress continues, it sees this number rise.

But quite apart from the problem of choosing primary qualities for mathematical treatment, the adoption of mathematical language raises the further problem of approximation in measurement. To apply mathematics you have to have scales, and scales like rulers have only limited accuracy, so that scientific data are never absolutely accurate. In the course of their training, scientists are generally taught to cope with experimental error, and not to quote results to higher accuracies than the experimental data will justify. However—and the point is one of Duhem's many claims to originality in this book—approximation in measurement has deeper consequences than the mere need to avoid spurious accuracy. Duhem broaches the issue in his third chapter. His example relates to the classic problem of the stability of the solar system.[33] As Duhem surely was aware, the problem was of some ideological importance. Newton had believed the solar system unstable—God had to intervene from time to time to put things right—but believing that he had proved the solar system stable as described by Newton's laws, the freethinker Laplace felt able to tell Napoleon "Je n'ai pas besoin de cette hypothèse" (I have no need of that hypothesis [of God]). In the late nineteenth century the researches of Henri Poincaré had re-opened the problem Laplace thought he had solved, and in the ensuing debate Jacques Hadamard applied the methods of the new mathematical science of topology to show that on his model at least, it could be quite insoluble, a result which had been discussed in correspondence between these two friends. Duhem comments on its significance:[34]

For the mathematician, the problem of the stability of the solar system certainly has a meaning, for the initial positions of the planets and their initial velocities are, for him, factors known with mathematical preci-

> sion. But, for the astronomer, these factors are only determined by physical procedures. These procedures involve errors, errors the improvements made to the instruments and methods of observation more and more reduce but never remove. Hence, for the astronomer, it could turn out that the problem of the stability of the solar system was devoid of all meaning; the practical data supplied by him to the mathematician are for the latter equivalent to an infinity of theoretical data bordering on each other but nonetheless distinct. Perhaps there are among these data some that would keep the planets eternally at finite distances, while others would send one of these celestial bodies away into the depths of space.

That is, insoluble in practice, the problem could "lose all meaning *for physicists*". Moreover, Duhem hints, this problem may not be the only one in mathematical physics in this condition.

6. Theory and Experiment

Duhem's analysis of the relation of theory and experiment occupies, in one form or another, the next three chapters, on 'Physical Experiment', 'Physical Law', and 'Physical Theory and Experiment'. Their main message is well known: the irreducibly theoretical component of the supposedly empirical aspect of physical science. In physics, as Duhem conceives it, there is no trace of the simplicities of direct observation; you have to study electricity to understand an experiment in that subject[35] and your naive untrained observational powers are of no use here. Roberto Maiocchi sees this as primarily directed against the instrumentalism of Duhem's day.[36] It will be recalled that instrumentalist philosophies of science deny to scientific theory the power of saying anything about how nature truly is, generally conceding to it no more than the power of summarizing observational information, whereas realist theories insist that the truth about nature, as it really is, is and should be the aim of science. According to Maiocchi, instrumentalism depended for its plausibility on assuming the availability of theory-free observational knowledge, knowledge whose recording and discovery could be said to be the prime purpose of physics. But the effect of Duhem's analysis is to destroy that assumption.

If Maiocchi is right, and I find his case plausible, then the point of these chapters is *not* an attack on the foundations of theory, but on the reliability of facts. Thirty years later, Karl Popper was equally to deny the possibility of a theory-free empirical basis for science, and in a phrase which admirably captures the spirit of Duhem's analysis, to say of theory that it consisted of "piles driven into the swamp".[37] For both Duhem and Popper, the critical analysis of experiment has the effect of asserting the rights of theory. For the positivistic instrumentalists it was the facts that could be relied on, and theory only existed on sufferance. For Duhem and Popper the boot is largely on the other foot. As is well known, of course, Popper goes on to attempt a rehabilitation of truth as the goal of theorizing. Whether Maiocchi is right in thinking that Duhem likewise went on from the rehabilitation of theory to the rehabilitation of truth is difficult to say: it all depends on how we interpret the doctrine of natural classification discussed above. If it is read as a theory of partial truth, or verisimilitude, then Maiocchi's claim has to be accepted, but it is a weak form of partial truth, and it seems best to leave this question here, though a later chapter will offer an interpretation of *To Save the Phenomena* that assumes that Duhem subscribed to a qualified realism.

In any case, the bearing of Duhem's critique of experiment on the question of a realistic or instrumentalistic interpretation of physical theory is not my prime concern, nor is it the aspect that has attracted the attention of philosophers of science: that has been rather the effect of the theoreticity of facts on the experimental testing of physical theories. This was certainly also an issue for Duhem, for much as he believed in the rights of theory, he also believed in the importance of experiment.

But if experimental facts and laws need theories even to state them, then the testing of a theory looks like a circular exercise in which at best one theory is being compared with other theories; for many readers of Duhem this result has been at least a challenge, if not a scandal, and it posed a serious threat to the various empiricist philosophies of science that flourished in the middle years of this century. As such it has been discussed *ad nauseam* in the literature of the philosophy of science, and the debate continues. Many of the main contributions to the debate on the so-called 'Duhem-Quine thesis' that has arisen out of it are conveniently available in the volume edited by Harding.[38] I do not propose here to 'answer' either Duhem or his critics, but to make a few limited points that may help put the debate in perspective.

Firstly, if Duhem's argument is to be understood, attention must be paid to the last section of Chapter vi. Duhem, readers will not now be surprised to learn, meant what he said about *bon sens* or intuitive judgement being the guide to which hypotheses have to be abandoned, for it was the intended conclusion of the argument of the three chapters, and *bon sens* includes judgements about whether alternative hypotheses are to be rejected as trivial or absurd. For Duhem "logic is not the sole guide of our judgements." "How," asks Grunbaum,[39] "does Duhem propose to assure that there exists . . . a non-trivial set A′ [of alternative auxiliary hypotheses] for any one component hypothesis H [of a theory] independently of the domain of empirical science to which H pertains?" Duhem assured or proposed no such thing: he was talking about the limitations of deductive logic, and triviality is a matter of judgement not logic. Similarly, to Popper's claim[40] that Duhem's argument ignores the fact that in practice we can easily solve the problem of what hypothesis to abandon the answer is that of course we can (or often can), and, as I shall argue below, Duhem knew this perfectly well—for him, as it must be for someone who takes his logic seriously, it was a matter of judgement, not logic. "Logic is not the sole guide of our judgements."

The example already mentioned above of Wiener's experiment on the relation of the plane of vibration to that of polarization in the vibratory or wave theory of light is instructive in this regard, for it shows us the kind of thing Duhem meant.[41] The theory in question may be inconsistent, but there may be a way out and it is difficult to determine whether there really is such an escape route. Wiener's experiment, said Duhem, had cut down the options. It seemed unreasonable to adopt Poincaré's suggestion that we abandon the assumption that the intensity of a light beam is proportional to the square of the illumination. If we reject this and others like it, we have to give up the elastic theory of light. There are no logical grounds to enforce the conclusion that the elastic theory is falsified, but "logic is not the sole guide of our judgements".

Secondly, attention should be paid to its place in the overall structure of the book: as the counterpole to Duhem's earlier argument that "No metaphysical system is enough to erect a physical theory".[42] If metaphysics cannot do the job, perhaps, it might be thought, experiment has more prospect of success. After all, Duhem has told us, experiment is the sole criterion of the truth of a physical theory.[43] If such a criterion is worth anything, it ought to be able to get us some truth, and presumably also tell us something of the nature of the world as well, thus getting us out of the dilemma Duhem posed at the beginning of that chapter, between giving up explanation, and submitting to the chaos resulting from the subordination of physics to metaphysics. What Duhem is now doing is cutting off that escape route. Even here, the attempt to use a *géométrie* founded on experiment to deduce the hypotheses of physical theory fails, and leaves us alone with *bon sens,* with physical laws that are provisional and relative because they are approximate and because they are symbolic, and with theories that are equally so. Neither metaphysics nor experiment can supply us with hard *criteria* of truth in physics.

Duhem looks at each of the traditional methods of establishing a *géométrie* of experiment in turn. Falsification fails because any experimental test involves a system of hypotheses, not just the one at issue,[44] and *a fortiori* the attempt to turn hypotheses into demonstrated truths by the *experimentum crucis* (crucial experiment) of Newton also fails,[45] for the fact that it is always a *theoretical system* that is at issue ensures that in principle there is always an escape route capable of saving, for instance, the Newtonian emission system of optics from Foucault's purported experimental demonstration of the wave theory. Finally, the method of deduction from the phenomena Newton codified in his *Regulae Philosophandi* (Rules of Philosophizing) collapses on close analysis.[46] Newton did not, and could not, have *deduced* universal gravitation from Kepler's laws of planetary motion. To get his result he had to generalize and correct, neither trivial processes, for Newton's law is actually inconsistent with Kepler's laws, and it is only preferable to the latter because the difference between them can be used to calculate the *perturbations,* the departures from the latter of the planetary motions actually observed. Duhem winds up with a similar critique of Ampère's application of the Newtonian method of deduction from the phenomena to his 'deduction' of the laws of electrodynamics.

Thirdly, a point I shall return to in Chapter X, is the extent to which the additions to this analysis reflect the way in which the original journal article on which it was based had become part of an ongoing argument that was still very much alive when Duhem was writing. He had to remind his readers of his priority over Milhaud, Wilbois, and le Roy, and he uses the fact of this priority to pick up on Poincaré who, while arguing against his positions, chose not to cite him, naming le Roy instead.[47] One of Duhem's subsidiary aims in this part of the *Théorie Physique* seems to have been to maintain his position in an argument that he had helped to start, while avoiding getting involved in the Bergsonian positions of such as le Roy, "who goes beyond the bounds of physics".

Finally, needless to remark, it is a classic and deserves to be. The reason why is not hard to find: it is a really critical analysis by a first class mind with the necessary imagination, who knew the subject intimately, and had the independence of mind not to be hypnotized by received opinions. As he said in his 'Introduction':[48]

> Moreover, the doctrine set out in this work is not a logical system resulting solely from the contemplation of general ideas, nor constructed by a meditation repelled by concrete detail. It was born and was developed by the daily practice of the Science.

There are not many works in the philosophy of science of which that could truthfully be said.

7. History and the Non-Choice of Hypotheses

The final chapter, 'The Choice of Hypotheses', is the culmination of the book. If I am right, it will leave intuitive judgement in complete charge, and in a way this is what it does. The conditions imposed by logic on the choice of hypotheses are, as Duhem has said before in 1892, and as he now repeats,[49] minimal, but, against Vicaire, who complained that Duhem's 1892 theory gave no guidance on the matter,[50] the problem of the choice of hypotheses is a non-problem, for hypotheses are the result of a continuous evolution and just emerge.[51] Consistently with this view, he recommends a historical approach in teaching, both because it is better than presenting hypotheses by the teacher's *fiat* before they have been understood, and because of its educational value in helping students to become familiar with the subtle relationship between theory and experiment.[52] In passing he has a dig, referred to above, at Euler's proposal that physical theories should be deduced from axioms supplied by common knowledge.[53]

Most of the themes already noted in its predecessors reappear in this chapter, though here there is rather more emphasis on intuition than on deduction, but there is the same repetition, the same tendency to return to the same points from different angles. Thus, for example, Section iv, 'On the Presentation of Hypotheses in the Teaching of Physics', covers much the same ground as Section vi of his previous chapter, 'Consequences relating to the teaching of Physics', though it takes the issues raised rather further, and arrives at a conclusion concerning the dogmatic presentation of hypotheses, which is entirely consonant with Duhem's general thesis, and obvious anyway to anyone who has ever thought about teaching.[54]

> It is obvious that that style of expounding Physics, the only one that would be perfectly logical, is absolutely impracticable. Hence it is certain that there is no logically absolutely impeccable form in which Physics teaching can be carried out: *every exposition of physical theories will necessarily be a compromise between the demands of logic and the intellectual needs of the student.*

The theme of the tension between deductive rigour and intuition is thus reiterated in a way that recalls the Pascal of the *Esprit Géométrique* and the Duhem of 1892.

Duhem's explicit refusal to countenance attempts to deduce physical theory from allegedly commonsense axioms has already been alluded to several times. It is hardly necessary to go into the reasons why he judges to be fraudulent Euler's attempt to deduce Newtonian mechanics from such principles: my concern is more with the occasion it provides Duhem to spell out the tension between intuition and deduction in the structure and progress of physics, and the consequences he draws from it. Duhem reminds his readers of his belief that *common sense* is the source of the principles of mathematics:[55]

The idea of demonstrating from commonsense knowledge the hypotheses physical theories rest on is motivated by the wish to construct Physics on the model of Geometry. Indeed, the axioms Geometry is derived from with such perfect rigour, the postulates Euclid formulates at the beginning of his *Elements,* are propositions whose self-evident truth common sense affirms. But we have seen, several times over, how dangerous it is to conflate the method of mathematics with the method followed by physical theories . . . The mathematical sciences are very exceptional sciences: they alone have the blessing of treating ideas arising from our daily perceptions by a spontaneous work of abstraction and generalization, and yet they subsequently show themselves neat, pure, and simple.

Physics is refused this blessing: the notions . . . it has to treat are . . . infinitely confused and complex, notions whose study demands a long and painful job of analysis . . . ; the men of genius who created theoretical Physics understood that to get order and clarity into this task, these qualities had to be demanded of the only sciences naturally ordered and clear, the mathematical sciences; but they were not able to bring it about that clarity and order in Physics could, as in Arithmetic and Geometry, be combined in any direct way with obvious certainty.

Like the teaching of it to students, mathematical physics is thus a compromise, a compromise between the demands of logic and, this time, not the needs of the student, but those of common sense:[56]

In the domain of observational laws, common sense reigns: it alone, by our natural means of perceiving, and judging our perceptions, decides on truth and falsity. In the domain of schematic representation, mathematical deduction is the sovereign mistress: all must parade to the rules she imposes. But from one domain to the other there is established a continual circulation, a continual exchange of ideas . . . In the intermediate zone across which these exchanges take place, . . . common sense and mathematical logic make their influences compete, and intermingle inextricably the procedures that belong to them.

Physics, then, depends on this intermediate zone where theory keeps calling for new experiments, and observation suggests amendment to the theories in use. Duhem comments that mathematicians are liable to forget the existence of this zone and attempt to construct their physics like their mathematics, with results "Ernst Mach so rightly calls *false rigour*".[57] Which of course is to leave a problem: how do you train physicists so that they understand this situation, so that they understand the continual exchange of ideas between theory and experiment, so that they understand, he might have said, the necessity of this tension between logical rigour and intuition? Duhem's answer, of course, is 'through history', an answer prepared long in advance by the use of historical matter earlier in the book and, prominently indeed, earlier in this chapter. Duhem puts his point of view in a classic metaphor: just as, as he has said earlier, theories are not devised or imagined by their inventors in one piece from nothing, that is not needed here either:[58]

> But as to that method, why try to imagine it complete? Don't we have in front of us a student who, entirely ignorant of physical theory in his childhood, has come as an adult into full knowledge of all the hypotheses these theories rest on? This student . . . is humanity. Why, in the intellectual formation of every man, do we not imitate the progress by which human science was formed? Why do we not prepare for the entry of every hypothesis into our teaching by a summary but faithful exposition of the vicissitudes that preceded its entry into the science?

History, it will be recalled from an earlier chapter, is the study to which Duhem denied any kind of *géométrie*, any kind of deductive method. This invitation to adopt a historical approach emphasizes the importance of non-deductive *finesse* for Duhem, and invites the reader to begin consideration of the historical work for which he is famous. To it I now turn.

VII

Critical History and Its Assumptions

1. A Historical Example and Its Rôle

This and the following chapters will explore the consequences, expected and unexpected, of Duhem's commitment to the historical method, "the legitimate, sure, and fertile method of preparing a mind to accept a physical hypothesis." The final chapter of his *Théorie Physique* is as good a place as any to begin. By a short head the second longest in the book, its length is largely due to the 52 pages of Section ii that use the example of the history of universal attraction from Aristotle to Newton to argue the thesis that "hypotheses are not the product of sudden creation but the result of progressive evolution".[1] In the face of the extreme freedom that logic leaves the theorist, Duhem has put the rhetorical questions: 'Can man advantageously make use of such unlimited freedom? Is his intelligence powerful enough to create a physical theory in its entirety?' His answer is 'Certainly not'. In Duhem's view the story of universal attraction shows how ideas gradually developed to the point at which Newton's system became possible, how Newton emphatically did not create his theory of universal gravitation from nothing, but as he famously said, "stood on the shoulders of giants".

Duhem also uses history much more briefly to illustrate a second, and more questionable point, the thesis that there is really no question of choice at all: "The physicist does not choose the hypotheses he will base his theory on: they germinate in him, without his assistance".[2] Here, the history is much shorter, with various examples from modern physics, including a couple of simultaneous discoveries: ideas appear when their time has come. He gives the examples of the simultaneous discovery of universal gravitation in the minds of Hooke, Wren, Halley, and Newton, and of the equivalent of heat and work in the minds of Robert Mayer, Joule, and Colding, using the second example for very different purposes from those of T. S. Kuhn.[3]

The material referred to above is all, of course, an expression of the famous continuity doctrine, which, for reasons that are not entirely clear, has come to be regarded as controversial. Here, I have little to add to the extensive and interesting discussions of Maiocchi,[4] and of Ariew and Barker.[5] The former sees it as basic to Duhem's methodology from the beginning of his career, and the latter see it as evidence of realism. I have no objection to either interpretation of a principle that to me seems just obvious. Duhem never held that changes of direction in the history of physics did not or should not occur—deciding when was as always for him a matter of intuitive judgement—though he did demand respect for tradition. In this he seems to be self-evidently right, and philosophers of the school of Popper ought to be acknowledging their basic agreement with him. After all, a basic principle of Popper's social philosophy is the rejection, on the grounds of irrationality, of what he calls utopian social engineering, of the attempt to reconstruct social systems completely from the beginning on *a priori* principles leaving nothing unchanged;[6] total theoretical change is the epistemological equivalent of that enterprise. Be that as it may, at one level, the continuity thesis seems to express no more than trite common observations, and due humility. Just as Newton acknowledged that he stood on the shoulders of giants, Pascal expressed his debt to the ancients in his *Préface sur le Traité du Vide*.[7]

> . . . today we can have other thoughts and new opinions without disregard or ingratitude, . . . because, after getting up to the one level they have carried us to, the least effort makes us rise higher, and with less trouble and less glory we are above them.

In passing, it is worth noting the language of germination Duhem habitually uses to express his continuity doctrine. Its appropriateness may well be debatable: the idea it expresses that there is no element of choice in the growth of science may, perhaps rightly, be felt rather far-fetched and objectionable. However, it does seem to point to a feature of the experience of many of us, that we find ourselves often quite uncertain, not only of the source of our ideas, but uncertain even about when we first had them. It is also one more piece of evidence linking him with Louis Pasteur, whose laboratory was still based at the École Normale in the mid-1880s when Duhem was a student there. It is well known that, in the course of a long polemic against theories of spontaneous generation Pasteur established the modern view that fermentation and disease were the work of micro-organisms or 'germs'.[8] A statement Duhem published the year before in the *Origines de la Statique* is so redolent in its language of the spontaneous generation controversy, that the allusion can hardly be other than deliberate:[9]

> Science knows nothing of spontaneous generation: the most unexpected discoveries were never created entire in bosom of the intellect that gave them birth; they are always the issue of an initial germ first deposited in this genius; his rôle is limited to increasing and developing this little seed sown in him till the tree with its powerful branches gives its flowers and its fruits.

It has to be said though, that, however expressed, the continuity thesis will not get students very far in understanding Duhem's historical output. It is often assumed that the continuity thesis is enough to explain his interest in mediaeval science. It most emphatically is not. As Maiocchi has shown, it was part of the conventional wisdom of positivists who showed no interest whatever in mediaeval science, and that statement includes the young Pierre Duhem. If his historical work is to be understood, clues to it have to be sought elsewhere, and the continuity thesis, which seems to dominate so much writing about Duhem, will play no further rôle in what follows. It is irrelevant.

2. History for the Purposes of Physics

The historical method, claimed Duhem,[10] is the legitimate, sure, and fertile method of preparing the mind to understand and accept a physical hypothesis. If that is so, it ought also to be the legitimate, sure, and fertile method of preparing the minds of Duhem's readers to understand and accept his account of physics and of physical method. It is now time to examine the kind of historical writing that resulted. As I have already suggested, the picture I shall present is by no means the expected one. The outstanding fact to notice at the outset is that the work that issued from Duhem's historical interests itself poses a historical problem because of the extent to which it changed in character in the course of his career.

To anyone familiar only with Duhem's later work on mediaeval science, the early work must come as a startling contrast: instead of multi-volume works running to thousands of pages, we have brief, brilliant, summary expositions of a couple of hundred; instead of works whose prime focus is on the scientific achievements of the Middle Ages, we have expositions of the scientific achievements of Pierre Duhem, expositions in which there is not a trace of the Middle Ages, except by way of denigration; instead of the wide-ranging account of the *Système du Monde*, in which the science is integrated into its wider philosophical and theological context, here Duhem's story is confined to the sciences in question, and is narrowly *internalist*, altogether separated from all non-scientific, or *external* consider-ations. This contrast is well-suited to bring out Maiocchi's claim that positivism provided the essential context of Duhem's work; I have remarked on it several times already in the course of this essay, and it is now time for a fuller account of it and an attempt at a systematic examination of the principles and development of Duhem's historical work.

History of science of the kind he was calling for was nothing new: it was on the contrary an established *genre* of the time, of which the best-known example is perhaps the *Science of Mechanics* of Duhem's older contemporary and correspondent, the Austrian physicist and philosopher Ernst Mach (1838–1916), whose full title, *Die Mechanik in ihrer Entwicklung, historisch-kritisch dargestellt* (1887: Mechanics in its Development, a historical-critical account; commonly known by its English title *The Science of Mechanics*), gives an adequate idea of its character. Mach advocated a type of positivist epistemology in which sense experience and nothing else was to be the basis of all knowledge, whether of physical or of mental things. For this theory of knowledge, theoretical constructs like Newton's forces were an embarrassment, and in the *Science of Mechanics* he mounted a quasi-historical critique to argue that they were logically unnecessary and so eliminable. In the treatment of his material that aim necessitated a close critical engagement of the kind met with in much history of philosophy practised by philosophers, in the *Kritische Geschichte der allgemeinen Principien der Mechanik* (Critical History of the General Principles of Mechanics) of Eugen Dühring (1833–1921), and very visible indeed in the work of Duhem himself, right to the end of his career.

Duhem's own historical writing began quite early. The accepted thesis on magnetization by induction of 1888 was accompanied by a survey of the history of the same subject.[11] I have already referred to his 'Notation Atomique' of 1892, the article that was to be the harbinger of many more like it in the 1890s, single-part or multi-part journal articles on topics connected with his interests in physics, of which a notable example is the three-part 'Les Théories de la Chaleur' of 1896, published in the *Revue des Deux Mondes*. At the end of the decade, however, both the tempo and the scale of Duhem's historical writing increased. Extended multi-part articles appeared, which were then brought out as brilliant monographs: *Les Théories Electriques de J. Clerk Maxwell* (1900); *Le Mixte et La Combinaison Chimique* (1900–01); and *L'Évolution de la Mécanique* (1903). Each of these had a precise polemical point within the debates of the time: outrageously so in the case of the first, which claimed that the electromagnetic theory of Maxwell (1831–79) was a confused mass of error and bad practice, for which the only cure was the alternative theory of Hermann von Helmholtz (1821–94); but equally so in the case of the other two, which both represent historical propaganda for Duhem's version of the energetic programme, the reduction of the whole of physics and physical chemistry to the laws of energy. In both the pattern is the same: the first two-thirds of the book presents an abbreviated historical account of the rise and fall of previous theories, while the last third presents Duhem's alternative and its supposed advantages. The *Évolution de la Mécanique* comes last in time in this group: for this and other reasons, it will be subjected to closer examination.

3. The Evolution of Mechanics

The subject-matter of mechanics is the motions of bodies; how fast they go, their collisions, and the forces that make them go faster or slower—on the face of it a relatively simple subject that should not require too much theoretical sophistication, but in practice raising difficult problems that proved to be none too easy to solve. The title of Duhem's work might lead the casual reader to expect a technical history of the theories of mechanics as such, of the principles of the subject and of the processes by which these principles came to be embodied and applied in the theory. These concerns are not absent from Duhem's exposition, though they are present in a rather non-technical way, dictated no doubt by the largely non-technical readership of the *Revue Générale des Sciences* in which it was first published.

It soon emerges, however, that Duhem had wider concerns in view: mechanical explanation in physics, and the rôle of mechanics in the overall scheme of that subject. On reflection, this widening of the field of inquiry is unsurprising, and that for two reasons: *via* its mathematical expressions for such concepts as motion, acceleration, force, energy and the like, mechanics supplies the essential apparatus for the theory of every other part of physics; and as Duhem was only too well aware, a constant feature of the history of modern physics going back to the seventeenth century has been persistent attempts to reduce every other part of physics to mechanics and nothing else, attempts mostly involving the supposition that the matter of the bodies about us is divided into tiny invisible atoms. Duhem, on the other hand, claimed that his own work in physics offered an approach to physical theory that made all such atomistic and mechanistic suppositions unnecessary, one that supposed nothing but suitable mathematical descriptions of the things we could see and touch. A rapid survey of the book's contents will give some idea of its general character.

Duhem opens with a summary presentation of the main seventeenth-century positions, a presentation that seems to be heavily dependent on Arthur Hannequin's account[12] of the history and problems of atomism. Where Aristotle offered a qualitative account of the world based on the qualities of objects as they appeared to us, none being given any special status or privileges over any of the others, the figures who created the seventeenth-century scientific revolution took a different road, giving a privileged status, as so-called 'primary' qualities, to those mathematizable properties of objects considered solely as mobile things; they proposed to explain all observable phenomena by the laws of mechanics, and in this they did not restrict themselves to physics: chemistry and living things also came within their purview. They would all have agreed with the celebrated Dutch physicist Christiaan Huygens (1629–95) that in the true philosophy all the appearances of the world should be explainable by mechanical reasons if there were to be any hope of understanding them at all; there was, however, some dispute among the supporters of this 'mechanical philosophy' about which qualities were primary and how many of these there were.

Following René Descartes (1596–1650), there were the *Cartesians* who claimed to explain everything using nothing but the single category of extension, or the mere occupation of space. Shape and motion were to suffice for their physics, and they dismissed anything else as an occult quality, one of those obscure 'virtues' of Aristotelian or Scholastic origin they hoped to banish from science. Their system had problems. Consistently with their principles, their version of the 'mechanical philosophy' had no room for notions of empty space, for a void: space *was* matter. But in that case how could they give separate identities to the particles they ascribed shapes to, and whose collisions were supposed to explain observable phenomena? To *atomists* like Pierre Gassendi (1592–1655) and Huygens the Cartesian programme seemed impossibly restrictive. Following the ancient precedent of Epicureans such as Lucretius (99–55 B.C.), the physics of the atomists proposed that the world consisted of nothing but atoms moving in the void. Their atoms were allowed the property of mass in addition to extension or the mere occupation of space. On that basis, Huygens set out to explain other properties like weight. To others, like Isaac Newton (1642–1727), even this was too narrow: matter needed to be allowed attractive and repulsive forces as well if the programme of mechanical explanation was to be viable. To the most significant of his rivals and contemporaries, Newton's move seemed insupportably retrograde: universal attraction was no better than an Aristotelian occult quality. But Newton's emphasis on the need for forces in physics was in harmony with the central claim of one of the most important of his critics, G. W. Leibniz (1646–1716), that a mechanics with no room for force could not suffice: what was needed was a *dynamics* that had room for the active powers of matter.

But whatever is permissible in the fundamental principles, a programme of mechanical explanation requires a formulation of mechanics that will serve its needs: the increasing mathematical sophistiction of the eighteenth century encouraged the development of rational mechanics, the highly abstract mathematical presentations of mechanics that underlay virtually the whole of nineteenth-century physics, and are still prominent in advanced work in the subject. Of this all that is relevant here are the fundamental aims lying behind the approaches adopted. J. L. Lagrange (1736–1813) offered an abstract, mathematical theory, an *analytical mechanics*, capable of handling the interactions of any real bodies of the sizes normally met with, irrespective of any suppositions that might be made about the constitution and properties of these bodies. Denis Poisson (1781–1840), on the other hand, rejected all such approaches as false to nature. He aimed at a *physical* mechanics, one in which the mechanical properties of ordinary bodies were to be explained as resulting from the atoms, the so-called point masses Poisson's physics told him they were made up of.

As Duhem saw it, bold and ingenious as Poisson's approach was it soon revealed its limitations, and in fact the Lagrangian analytical approach came to underly the work both of atomists like James Clerk Maxwell, and of anti-atomists like Duhem. The difference lay in the use they put it to: supposing the world and everything in it to consist of the mechanical systems they could apply Lagrangian mechanics to, Maxwell and his disciples were compelled to suppose a plethora of hidden motions of which there could never, it seemed, be any evidence, while Duhem in contrast proposed to extend Lagrangian mechanics by including additional observable large scale properties in his theory. Where, for example, Maxwell and his disciples went to elaborate lengths to explain atomistically why cool bodies did not become hot unless heated, Duhem proposed to take it as a fundamental principle—the so-called Second Law of Thermodynamics—that they just never do it, and use this principle to rationalize other observable phenomena.

Duhem's work is marked by a triumphalist style in dealing with the progress of the science, of the kind usually dubbed *Whiggishness*, which makes 'what we now know' the norm for judging the past, and by a frequent critical tendentiousness, but its analysis is often acute and illuminating—beginning students could do worse than read this brilliant essay before proceeding to other more mature historical works. However, its most obvious feature for modern readers is the total absence of anything to do with the Middle Ages: the opening chapter concerns Aristotle, and the second Descartes: the ages in between are passed over in silence.

The fact must surely astonish those who know Duhem only by his reputation as a historian of mediaeval science, particularly when they recall that Duhem died in 1916 and that this was published in 1903, when he had only thirteen years left to live. One may be inclined to grasp at the straw that at least Aristotle is taken seriously, but still this work seems to be written on the assumption that the Middle Ages added nothing whatever to what Aristotle wrote. In this the *Évolution de la Mécanique* is, if anything, an improvement on the other productions of Duhem's earlier years: his May 1894 *Revue des deux Mondes* article, 'Les Théories de l'Optique', starts with Descartes and assumes that there was nothing interesting before then. As I have remarked already, Duhem might well have believed in the continuous development of physical theory, but he does not seem to have thought himself thereby required to search for a physics in the Middle Ages. Equally, there might well be a Scholastic revival going on when Duhem was writing, and as a Catholic he must certainly have been aware of it, but the evidence is unmistakable: *for the first twenty years of his adult career, Duhem regarded the Middle Ages as scientifically a nullity, and in this he shared the assumptions of his positivist contemporaries*. That being so, neither Duhem's continuism, nor any kind of scholastic orientation deriving from his Catholicism can explain his later interest in mediaeval science. The most that might be said is that these factors made that discovery more palatable to him when he was in a position to make it.

When Duhem ignored the Middle Ages in this way, he was doing no more than following the example of his positivist models such as Mach and Dühring, for whose historical works mediaeval science did not exist either. Dühring, indeed, said as much in his preface: between Archimedes and the sixteenth century there was nothing, only a 'historical desert'.[13] Assertions of this sort were in any case part of the positivist anti-religious case. For were not Western education and culture then under Church control, and did not the Church oppose the development of science? How could the positivists do other than assume the scientific nullity of the Middle Ages unless they were going to subject their views on these matters to radical revision? The surprising thing is that at this point in his career, Duhem the Catholic accepted their view of mediaeval science as fact. My next chapter will explore how Duhem came to change his mind, but for the present, something has to be said about the style of the critical analysis adopted in the earlier works.

4. The Aims and Norms of Historical Analysis

I have already remarked on the close critical engagement with his material involved in the critico-historical style Duhem shared with his positivist predecessors. This style raises the interesting question of the precise relation between Duhem's historical work and his methodological principles. The two were intimately connected in his mind, since, in his conception, physics was to be an application of his methodology. The connection between Duhem's theory of physical inquiry and his historical practice has long commanded the attention of scholars: no account of Duhem's philosophy of science is complete without it.

It is a commonplace of historiographical discussion, and obvious anyway to anyone who thinks about the matter, that no history can ever be complete. The information available to historians is unlimited. If they are to tell us anything at all they must select. When they select they must of course not mislead their readers: they must, that is, show the expected academic integrity; but fulfilling that requirement leaves a great deal of scope, depending on the historians' aims and their conception of their material, matters that are particularly tricky for historians of science, and for intellectual historians generally. It may, for example, be asked whether the purpose of the history is to be purely illustrative, or whether it is also to be critical. In the latter case it may be asked what standards of criticism are to be applied, and in any case, the consumer of history needs to know what principles of selection are operative: it is not, for example, enough to know that the intention is to write a history of late nineteenth-century physics, since a lot may depend on what it is proposed to regard as physics, and how that subject is separated, not only from other sciences, but from non-scientific subjects. Observant readers of the previous chapter, for example, will have noticed that Duhem's conception of physics includes material more usually regarded as belonging to chemistry. Such questions are still under debate.

Particularly if it is carried out by a philosophically conscious historian like Duhem, the special problem of critical history of science or philosophy is that it inevitably involves the philosophical and scientific principles of the historian. The late Imre Lakatos once argued[14] for the use of history as a testing ground for theories of scientific method, or, as he would have put it, for theories of scientific rationality, the test consisting essentially in their effects on how history of science is written, on its historiography. For the reasons mentioned above, he regarded all history of science as inevitably involving the philosophy of science of the writer, if only via the criteria used in selecting material for inclusion, though in most cases the philosophy of science would not be explicitly articulated, but would remain implicit in the presentation of the material. Theories, then, in Lakatos's scheme, were to be judged according to the fruitfulness of the account of the history of science they yielded. For rather understandable reasons, histories guided by *implicit* philosophies of science, by philosophies that had not been articulated by their authors, were not very amenable to the kind of analysis called for by Lakatos's programme: to be any use they needed to be made *explicit*, and to make clear what he meant, he proposed the idea of a 'rationally reconstructed' history, history of science in which only what was philosophically relevant was included, omitting the (philosophically trivial) concrete historical circumstances of that history. In his classic *Proofs and Refutations*,[15] he consigned the latter to the footnotes.

But that is to open up a hornet's nest, with only, Duhem might have insisted, the intuitive judgement of the historian to guide us. Whether the aim of the history is critical or purely illustrative, the unavoidable question is whether the result is genuine history or no more than a caricature: if its aim is critical, whether we are now involved in a vicious circle, with the philosophy so inextricably involved in the 'history' that it is now useless for testing a theory of scientific method. Is Lakatosian rational reconstruction, to use T. S. Kuhn's phrase, no more than "philosophy fabricating examples"?[16] If Kuhn's suspicion is correct, and, as will be seen, it applies as much to Duhem's early work as it does to the work of Lakatos's last years, illustration, and not critical history, is all we shall ever get. Moreover, the attempt at criticism can make historians so determined to discover whether the past writer was by their criteria right or wrong, rational or irrational, that they forget the difference between the period under study and their own time, forget the likely differences in assumptions and presuppositions between historians and historical subjects. Objections of this sort have often been urged against the work of Pierre Duhem.

5. Illustration and Criticism in Duhem's Historical Writing

Illustration seems to be the aim lying behind the last chapter of the *Théorie Physique*. History is to be used to help students and others understand the hypotheses now used, that is, in effect, to illustrate, by comparison and contrast, present-day theoretical approaches. That indeed was the aim Duhem made explicit in 1892, when, after publishing his initial essay 'Réflexions', he seems, as he tells us, to have decided that such an abstract presentation would be hard to understand and needed illustration, illustration he went on to provide in his 'Notation Atomique', published in the next issue of the *Revue des Questions Scientifiques*. In this article, he attempted to show, in an argument he was later to expand in *Le Mixte* that atomic hypotheses could be written out of the development of nineteenth-century chemistry, that atomic notation could be treated as a kind of technique for the mathematization of the empirical facts concerning chemical analogy and chemical substitution, in the same sort of way as the new branch of mathematics called topology provided an analysis of spatial relationships without any reference to the distances and angles between the objects it treated. The use of this quasi-mathematical technique was quite compatible with rejecting the real existence of atoms. The element of rational reconstruction in this approach should be obvious: belief in atoms is treated as an irrelevant historical circumstance that accompanied the development of the really valuable technique of atomic notation, a circumstance which a rational account of the essentials of the development of nineteenth-century chemistry could safely omit. An awful lot indeed is going to be left out of "the summary and faithful exposition",[17] of the vicissitudes preceding the appearance of the ideas Duhem is interested in conveying to his students.

In 1913, however, Duhem was to offer an alternative motive for his historical work, one that brought him much closer to Lakatos's programme: the criticism of his methodological theories. As he put it to the Académie des Sciences on the occasion of his candidature for non-resident membership:[18]

> Every abstract thought needs checking by the facts; every scientific theory calls for comparison with experience; our logical considerations on the right method for Physics cannot be soundly judged unless confronted with the teachings of history. We must now turn to the gathering up of these teachings.

Whereupon he launched out into a survey of his historical researches up to that date before, eleven pages later, pronouncing himself satisfied:[19]

> Now we have seen no essential principle to be derived from the desire to resolve the bodies we see and touch into imperceptible but more simple

> bodies, none whose object is to explain sensible motions by hidden motions . . . Hence the history of the development of Physics comes to confirm what we have been taught by logical analysis of the procedures employed in that science. From the one as well as from the other, we have gained increased confidence in the future fertility of the energetic method.

The critical process, with the testing against difficult cases that such an aim implies, is not very visible, or at best not very explicit, and many of Duhem's readers must have wondered whether this was really the point of, for example, the lengthy presentation and discussion of Mediaeval Aristotelianism in Volume v of the *Système du Monde* (not yet written, of course, in 1913); but at least it *is* relevant to such earlier works as *Les Théories Electriques de J. Clerk Maxwell, Étude Historique et Critique*, to *Le Mixte et la Combinaison Chimique, essai sur l'évolution d'une idée* and to *L'Évolution de la Mécanique,* though it is hard to see anywhere in the determined hunt for the precursors of modern science of the *Études sur Léonard de Vinci.* However, as remarked above, it shares with the aim of illustration the same acute problems of selection in dealing with historical data.

Duhem was quite explicit in the *Théorie Physique* about his selectivity:[20]

> Certainly it is not possible to retrace stage by stage the slow, hesitant, groping march by which the human mind attained a clear view of each physical principle . . . the evolution of each hypothesis must . . . be reduced in the proportion the length of a man's education bears to the length of the formation of the science; . . .
>
> . . . Besides, this abbreviation is almost always easy, provided we are willing to neglect everything that is merely accidental fact—author's name, date of invention, episode, or anecdote—in order to get at those historical facts only that seem essential in the eyes of the physicist, to get at those circumstances only in which theory was enriched by a new principle, in which it saw an obscurity vanish or an erroneous idea disappear.

This is surely the languge expected of a physicist, even if the physicist is a subtle one of wide culture. In the classic 'whiggish' manner, history is going to be written in terms of 'what we now know'. This expectation is both confirmed and disconfirmed by what he had to say about the work of that other believer in the historical method, Ernst Mach, whose admiration he very much reciprocated. In his 20-page essay review (1903) of the French translation of *The Science of Mechanics* Duhem commented on Mach's use of history. Emphasizing his own agreement with Mach about the importance of history, he notes:[21]

> If it is treated by the method Mr. Mach calls for, the history of Mechanics will appear infinitely interesting to the physicist, to him who seeks in the past only lights to lighten the present. If they forget that this is . . . the object the author wanted to attain, the historian and the psychologist will raise complaints against him.

They will, he says, complain about the omission of Descartes, they will complain about the excessive simplicity and order of Mach's story compared with the mess of reality, and they will complain about its excessive subjectivity, bearing too deeply the concerns of its author. Duhem's concluding comments show that even at this date the range of his concern is wider than might have appeared:[22]

> This last complaint, . . . will be occasioned by a reading of the chapter devoted to 'Theological, animistic, and mystical conceptions in Mechanics'.

Duhem made no comment at all on Mach's undoubted opposition to all such conceptions: he merely agreed with Mach that theoretical physics ought to be completely independent of any metaphysical or theological system. But he commented that though this was generally agreed when he was writing, that agreement had only recently been attained:

> . . . for many centuries Mechanics and Physics were most intimately linked to Metaphysics, to Theology, and even to the occult Sciences. This incessant action and reaction between the philosophical and theological sciences and Mechanics and Physics have to be constantly in the mind of anyone claiming to revive the ways of thinking of the creators of the Science . . . But, very often, when these laws [of natural philosophy] reached their definitive form, they showed themselves altogether severed from all the philosophical and theological ideas in whose bosom they long drew the nourishment needed for their development; . . .
>
> Hence, those who seek in the history of physical Science no more than a more complete knowledge of its material and concrete content, may almost always break the numerous links in that history with philosophical and theological systems. . .

I do not believe that this was merely politely worded criticism: Duhem was not known for softening his words to please, and he actually wrote that kind of history in the early part of his career. The kind of history that sought to revive the ways of thinking of the creators of physical science was all written after 1903, and all but a fraction of it after the *Théorie Physique*. The procedure he has just credited Mach with was his own: his was the history produced by a physicist whose only concern in the past was the search for lights suitable for lightening the present. It was, as I have remarked above, rational reconstruction, and it is not only in that respect that it resembles the work of Imre Lakatos.

6. Research Programmes Before Lakatos

The orientation of the *Évolution de la Mécanique* was thus towards a critical approach: it remains to note the manner of the criticism. Lakatos proposed[23] a model of scientific rationality in which the unit of evaluation was what he called the 'research programme', by which he meant, in slightly Pickwickian language, sequences of individual theories sharing common metaphysical assumptions and technical languages. Research programmes were said to be progressive if the successive theories in such a sequence were able to tell us more about the world, were of increasing content; otherwise they were said to be degenerating. In this work Duhem seems to proceed in a similar way, though he does so without the technical language that is the constant accompaniment of mid- to late-twentieth-century philosophy of science. Theoretical approaches are shown, in the manner of Lakatos's research programmes, as having striking initial successes before running into the sands of increasing contradiction and experimental difficulty. There is of course one difference. As an associate of Sir Karl Popper, Lakatos used Popper's criterion of empirical success: what the theory denies, what must be false if the theory is true. If, for example, we consider the convenient, though somewhat overworked, theory that all swans are white, the content (false as it happens) of that theory would be that there are no non-white swans, whereas Duhem is more content with facts his theory can interpret, even if it does not logically exclude other possibilities.

The extent of Duhem's agreement with Lakatos in his practical criteria for the evaluation of theoretical systems is unsurprising. According to Duhem, physical theory is to be autonomous, logically separate that is, from other subjects: hence if in real history it appears to be otherwise, a process resembling rational reconstruction will be required to separate out for criticism the hidden autonomous science that is really there all along, separating the essential from the accidental contaminants. According to Duhem also, experiment is to be the sole criterion of truth: that being so, theoretical ideas will inevitably be judged favourably if they are consistently successful experimentally, and judged unfavourably if they are constantly in need of repairs in the face of experimental anomalies. Lakatos's idea of the research programme articulated a basic intuition of practising physicists, an intuition likely to be shared by anyone who, like both Duhem and

Lakatos, has no confidence in the possibility of *attaining* certain truth. Hence, with his historical works up to 1903, Duhem showed himself in possession of historiographical practices in tune with his philosophy of science, and in their use he produced admirable specimens of the type. With both their brilliance and suggestiveness, as well as their frequent tendentiousness and triumphalism, Duhem's early works provide as good an illustration as any of the strengths and weaknesses of Lakatos's approach to the continuing problem of the relation between the history of science and its philosophy, to the analysis of the historiographical consequences of different philosophies. But all this was soon to become irrelevant. Ernst Mach was the recipient of a presentation copy of the *Évolution de la Mécanique*.[24] Within at most a month or two of his receiving it, its author was to make the chance discovery that would soon take him further and further from the concerns and assumptions that had given rise to that book and the earlier work that had preceded it, into situations in which, on Duhem's own admission, the crucial assumption, of an autonomous physics to be judged by experience alone, no longer applied.

VIII

The Discovery of Mediaeval Science

1. The Origins of Statics

The *Évolution de la Mécanique* assumed that as far as science is concerned the Middle Ages did not exist. *Les Origines de la Statique*, begun the following year, is the work in which this assumption is both explicitly stated and explicitly abandoned. For that reason it marks a decisive turning point in Duhem's career. Published in 1905–06, its two volumes are, in conformity with Duhem's typical publication style, apart from the prefaces, a verbatim reprint of a series of articles appearing in the *Revue des Questions Scientifiques* from the Autumn of 1903 to the end of 1906. His choice of topic is completely comprehensible to anyone familiar with his work, dependent as that was on the dynamics of Lagrange and the so-called principle of virtual velocities on which that rested, and was a natural follow-up to the *Évolution de la Mécanique*. The subject-matter of mechanics has traditionally been divided into two parts, *statics* or the theory of forces of equilibrium, and *dynamics* or the theory of how forces make bodies go faster or slower. Into the former go the laws of levers and pulleys, of screws and wedges, while into the latter go the laws of the motions of arrows and bullets, of planets and satellites. The principle of virtual velocities is a purely *static* one: it states that in a system of forces at equilibrium, any attempt to move the system away from the equilibrium point results in a proportionate restoring force tending to undo this action. However, Lagrange was able, in a technical manoeuvre discussed at length in the *Évolution de la Mécanique*[1] but here taken as read, to use it in combination with another principle, due to Jean d'Alembert (1717–83) and known by the latter's name, as the foundation of a dynamics, a science of the motion of bodies. Duhem had his reasons for finding this approach congenial: the science of heat, or thermo-dynamics, that he did so much to bring to its final form was a science of equilibria, of situations in which nothing is happening, and he too wanted to adapt such a science to serve as the basis for a theory of change: Lagrangian mechanics offered a model for the kind of extension he sought.

Nothing could be more natural than that the man who had published a book on the evolution of mechanics that gave such prominence to the Lagrangian approach should then undertake to explore the history of the static or equilibrium principles on which that approach was based. The reader can get a flavour of the work that might have resulted, but did not, from the first four chapters, published in the October 1903 issue of the *Revue des Questions Scientifiques,*[2] and reprinted without alteration in the first volume of the book. There is an opening chapter on Aristotle and Archimedes, followed by three on Leonardo da Vinci (1452–1519), a figure, it will be noted, who plays no part in the *Évolution de la Mécanique.* Chapter ii opens thus: "The commentaries of the scholastics on the *Μηχανικὰ Προβλήματα* [Mechanical Questions] of Aristotle added nothing essential to the ideas of the Stagirite."[3]

It is well to ponder this remarkable sentence. In the first place, there is the bland assumption that commentaries of no value were all that were written in the Middle Ages. In the second place, the attribution of the *Mechanical Questions* to Aristotle is doubtful. In the third, the work was altogether unknown in the Middle Ages.[4] Duhem can hardly have been doing any checking. He has made it only too clear that he knows nothing of the Middle Ages and does not care either. In a positivist, this would have occasioned no surprise, but in a Catholic? With a Scholastic revival going on all around him? Surely even the most peripheral acquaintance with neo-Scholasticism would have prevented this howler. But in October 1903 Duhem could not have been more remote from historical neo-Scholasticism, or from any knowledge whatever of mediaeval thought. There is nothing in this first instalment of *Origines de la Statique* to presage the historian of mediaeval science Duhem was to become, nothing to separate it in this respect from the *Évolution de la Mécanique*, and everything to separate it from the later chapters of the work of which it is part, for it is in the very next instalment, in Chapter v, that Duhem the mediaeval historian makes his first appearance: called 'The Alexandrian Sources of Mediaeval Statics' it opens in these terms:[5]

> Before coming to the fundamental treatise on statics produced in the Middle Ages by the enigmatic Jordanus de Nemore [c.1250], we must collect the debris scattered through the manuscripts of the writings composed in Alexandria on the science of equilibria.

There is no comment in the text to soften the contradiction, but it is no surprise that there was a gap in the publication series: the second instalment would have been expected in the January issue of this quarterly journal, but it did not appear till April. Fr. Henri Bosmans, author of an assessment of Duhem's historical work after his death in the *Revue des Questions Scientifiques*,[6] reports asking the then editor Jules Thirion S. J. for a sight of the manuscript, only to be met with the astonishing reply:[7] "I haven't got it yet. Duhem hasn't yet finished it. He says he still has some reading to do." To which Bosmans, who doubted the overwhelming importance Duhem was ascribing to Leonardo da Vinci, records his own comment: "Moreover I know two short treatises *On Weights*, both attributed to Jordan de Nemore. Duhem will eventually come across them, and I shall be surprised if he doesn't attach some importance to them." Indeed, Duhem seems to have been alerted to them by an incidental reference in a postcard from Paul Tannery[8] concerning a fragment on the science of weights attributed to Euclid. Bosmans knew of their existence before; perhaps he had even published on them; but his knowledge is irrelevant to the history of the rise of the study of mediaeval science: it was left to Duhem to use them to launch a new historiographical programme that remains controversial to this day.

The inconsistency in the published work is flagrant; Duhem made no attempt to smooth it away by careful editing; instead he wrote a preface to explain what had happened:[9]

> Before undertaking the study of the origins of statics, we read the few writings that treat the history of that science. It was easy to recognize that most of them were very condensed and lacking in detail, but we had no reason to suppose them incorrect, at least in broad outline. Hence, when we turned to the study of the texts they referred to, we anticipated having to add or alter many details, but nothing led us to suspect that the history of statics in its entirety would be upset by our researches.

Hardly the comment of a professional historian, who would not be expected to take the existing historiography for granted in this manner. But then Duhem was not a professional historian, but a physicist looking to the past for lights to lighten the present. For such a one, the existing historiography, if it is sufficiently detailed, will serve, for the interest lies, not in the details, but in the lessons for physics that the analysis of the physicist can draw from it. With statics, the problem that initially provoked Duhem's researches may have been that the existing secondary sources were over-condensed and insufficiently detailed to serve, but as Duhem was soon to discover, insufficient detail in historical accounts always means that the overall story is suspect too, that the material has not been sufficiently investigated to be sure of even the broad outlines. Duhem goes on to report a double surprise. In the first place,

Leonardo da Vinci was nowhere near as isolated as tradition made out: his work had been known and used by such important Renaissance mathematicians as Geronimo Cardano (1501–76) and Gianbattista Benedetti (1530–90). But with that correction the usual account still stood up. There was, however, another surprise in store:[10]

> We had begun to retrace this development in the hospitable pages of the *Revue des Questions Scientifiques*, when the reading of Tartaglia [Niccolò, 1500–57], whose name no history of statics even mentions, showed us unexpectedly that the work already begun had to be restarted on an entirely new plan.
>
> Tartaglia, indeed, long before Stevin [Simon, 1548–1620] and Galileo [Galileo Galilei, 1564–1642], had determined the apparent weight of a body on an inclined plane; he had derived it entirely correctly from the principle [virtual velocities] whose complete generality Descartes was later to affirm. But that discovery, mentioned by no historian of mechanics, was not his; in his work it was an impudent piece of plagiarism: Ferrari [Ludovico, 1522–1565] attacked him fiercely for it and claimed priority for a thirteenth-century mathematician, Jordanus Nemorarius.

2. Duhem's Sources

The modern physicist has perhaps taken too literally sixteenth-century litigiousness over intellectual property, but what is interesting about this passage are the reasons for Duhem's interest: the physicist has recognized correct physics: an important correct result is deduced from an important correct general principle, and the priority claim has forced his attention on to a period that, he has previously assumed, could safely be brushed aside. We are dealing with a physicist, with the attitudes of the physicist, including a strong interest in questions of priority. We are also, it must be said, dealing with a physicist whose knowledge of the secondary sources is very sketchy indeed, so much so that something must now be said about that literature and Duhem's use of it. Just what are the "histories of statics" that do not even mention the name of Tartaglia, and credit Stevin and Galileo with the first correct resolution of the inclined plane problem? They cannot include such standard sources as J. E. Montucla and Moritz Cantor[11] and the first instalment, of October 1903, does not cite them. On the other hand there are five references to volume iii of G. Libri's *Histoire des sciences mathématiques en Italie*, and one each to G. B. Venturi's *Essai sur les Ouvrages de Léonard de Vinci,* E. Wohlwill's 'Entdeckung des Beharrungsgesetzes',[12] and Part i, section i of J. L. Lagrange's *Mechanique Analitique* (*sic*). In footnotes, Libri[13] is criticized for careless and anachronistic reading of the texts, and Lagrange[14] for anti-Aristotelian prejudice. In view of the importance of his mechanics for Duhem's physics, Lagrange is a clear candidate for inclusion in Duhem's preliminary reading list, but Libri can be ruled out in view of the attention he gives Tartaglia, and Venturi seems likely reading only after Duhem had realized the importance of Leonardo. Lagrange on the other hand credited Simon Stevin with the first correct resolution of the inclined plane problem, and Guidobaldo del Monte (1546–1607) is the only other figure mentioned between Archimedes and Galileo.

If we look wider, the situation does not improve. In addition to Mach's *Science of Mechanics* Duhem would be likely to know of Düring's *Kritische Geschichte.* The former adds Leonardo to Guidobaldo and Stevin, the latter mentions Leonardo and Guidabaldo and refers to Libri and Venturi for information. Rühlmann's *Vorträge über Geschichte der theoretischen Maschinenlehre* adds little to the above. Such works, in the tradition of critical history discussed in Chapter VII above, seem to represent the type of books Duhem could have read. Their inadequacy as history is plain.

Inadequate as history they may have been, but something like them was what Duhem was writing, something governed by assumptions of the same order, merely more detailed.

After all, he was concerned, as he said, merely with the search for lights suitable for lightening the present. For such an enterprise, you don't need history, you don't need to know the author of the principle that interests you as a physicist, nor even the date of its discovery, only its content and how it grew out of earlier principles. For such an enterprise it ought to be irrelevant, actually, whether Tartaglia was really the inventor of the principle he used or a mere plagiarist, irrelevant whether anyone actually read Leonardo, irrelevant whether there was any science in the Middle Ages or not. In the 'history' Duhem had been professing, names and dates are mere local colour, not the substance. They are there to "add verisimilitude to a bald and unconvincing narrative".

So there we have it: by any standards an inadequate bibliographic base which, it is clear, Duhem intended to supplement by reading the primary sources referred to in the works it contained. He did not expect to have to rewrite the whole story, for that is not the sort of enterprise critical historians can be engaged in if they are to get on with their critical history: they have to be in a position to assume a standard history, a signposted road ready prepared for them to identify the texts they have to attend to. The problem for the student of Duhem is to understand how a critical historian came to regard his bibliographic base as inadequate, a problem, it may be remarked, not unlike that of understanding how a scientist comes to regard his factual basis as inadequate, and the theories he is working on as in need of revision. Duhem's bibliographic base was poor indeed and it was his own research that was to reveal its poverty, as first of all he busied himself with filling in the missing details, and then finally came to construct the coherent historical account which his own principles demanded. The question was, what that coherent historical account was going to be like, and how he was going to arrive at it. Was his road going to be like that of his neo-Scholastic contemporaries or was he going to side with the revisionists of his time who were finding a new Aquinas, different in many respects from the one presented in the Scholastic manuals? It turned out that Duhem's road was more the latter, definitely closer to that of the revisionists than to that of his scholastic contemporaries. Just as with his philosophy, his association with Blondel provides much of the essential topography of Duhem's historiographic road.

3. The Historiographical Muddle

In the meantime Duhem was in a muddle, a muddle with the dimensions of a crisis, and there can be no mistaking the signs: the man who in those very years had proclaimed such things as the date of a discovery and the name of the author irrelevant to the purposes of the physicist, was now engaged in a hectic hunt for the names and dates of Leonardo's precursors, and that on a scale and with an energy that continue to astonish. A principal cause of all this activity, it would seem, was that Duhem had found the existing historical framework, not just lacking in details, but altogether wanting, and without much further investigation the available primary sources just did not make sense. As he remarked in the preface, cited above, to the first published volume of the *Origines*, the two treatises attributed to Jordanus de Nemore were radically incompatible with each other, so that: "If we wanted to know just what mechanics owed to Jordanus and his disciples, we had to go to the contemporary sources, to the manuscripts".[15] A relatively limited task, one would have thought, but Duhem, it seems, thought otherwise: "There was no way out of analysing every manuscript relating to statics in the Bibliothèque Mazarine." Up to now, it might have been permissible to conclude that Duhem's enterprise was no more than damage repair, correcting the worst excesses in the received history so that his physicist's analysis could begin again on firmer foundations.

But on the contrary Duhem was now engaged on an altogether larger task, the discovery of the mediaeval roots of modern mechanics.

For Duhem's new historical investigations were soon to extend from statics into dynamics, from the science of weights to the science of moving bodies. In November 1903, he accepted Paul Tannery's invitation[16] to contribute a paper on some aspect of the history of mechanics to the history of science section Tannery was organizing at the Geneva international philosophy congress planned for September 1904. At that point he had no idea as to the subject. The manuscript that emerged was 'De l'Accélération Exercée par une Force Constant' and its theme was the mediaeval theory of impetus,[17] which has been much discussed since. The motion of projectiles had long been a weak point in the Aristotelian theory of motion. Part of a wider theory of any change whatsoever, this started from the obvious commonsense assumption that the natural state of things was no change at all, or in this case no motion at all. But it was a matter of common observation that arrows, crossbow bolts, boats, and mill wheels kept moving for a while after the initial push had ceased, and a variety of implausible theories were devised to explain this phenomenon. The modern answer is a theory of inertia, that the natural state of bodies is continued motion—that bodies keep going unless something stops them—and was the achievement of Galileo and Descartes in the seventeenth century. The latter of these also, as a measure of the force required to stop a body moving, multiplied the body's speed by its bulk to form what he called its quantity of motion, a definition later corrected, chiefly by Isaac Newton, to mass multiplied by velocity. Duhem's discovery was that three centuries before Descartes something like the law of inertia had been discussed in fourteenth-century Paris, where there was even an explanation for it, that a moving body had a force of movement, an *impetus*, to keep it going, a force that increased with its mass and with its velocity.

It was supposed that this force kept a projectile moving through the air once launched. This *theory of impetus*, which may have originated in Arab science or in late Antiquity, has played a large part in all subsequent discussions of the origins of modern dynamics. When Duhem delivered his manuscript in July 1904, his covering letter opened as follows:[18]

> I am sending you today the piece I promised you for the congress of the history of the sciences. It concerns the history of this proposition: *A constant force produces a uniformly accelerated motion*. When I took it up the subject was not entirely virgin. I hope, however, to have brought to light some poorly or wrongly understood facts.

Note well: the principal claim to originality is bringing to light some historical facts that had been poorly or wrongly understood: Duhem drew no lessons putatively beneficial to present day physics. A similar picture emerges from the fuller statement in Chapter v, in the April 1904 instalment of *Origines*:[19]

> But so far we have obtained only a crude sketch of the development of statics from antiquity to the Renaissance: to the essential outlines we have given, a mass of detail must be added.
>
> To establish this detail, we have had to impose laborious drudgery on ourselves: we have had to examine and analyse the many manuscripts relating to statics held in the Bibliothèque Nationale and the Bibliothèque Mazarine. This analysis has allowed us, we believe, to discover more than one spring, unknown or misunderstood until now, whose waters have copiously contributed to the formation of modern science . . .

Duhem had not only abandoned the doctrine of the sterility of the Middle Ages he had shared with his sources only nine months earlier, but with that abandonment there has gone, not indeed his critical attitude to his material, but certainly the historiographical programme he also shared with them, the pursuit of history of science solely as a guide to the present.

Duhem had discovered that his sources were so inadequate that he had to start again, and he was now discovering the facts, barrowloads of them. The discovery of facts, of the real history of mechanics before the sixteenth and seventeenth centuries, was now taking precedence over the the subtle analysis of the critical historian. The facts he looked for were, in the first place, those the physicist with a conscience about intellectual property would look for: the references omitted by the plagiarists who wanted to claim excessive originality for themselves, and so in the *Études* the facts are dredged up, out of context, so that the true originators of modern mechanics can be given their due honour, so that the true precursors of modern science can become known for what they are. But with that change of focus Duhem had done something else: he had also called in question the applicability of key parts of his theory of scientific method. Had he not said when reviewing Mach's *Science of Mechanics* that while everyone now agreed that physics ought to be autonomous and quite separate from metaphysical considerations, this recognition was only a very recent achievement: at earlier times physics and metaphysics had been quite inseparable.

How far Duhem realized this further consequence is hard to say: in the *Études* the focus is on what Duhem recognizes as the physics without regard to the intellectual and religious context in which the physics was bedded, but that he was prepared for wider considerations is clear even in the October 1903 instalment, at the close of Chapter i, when he compares Aristotle and Archimedes:[20]

> Hence, in the study of the equilibrium of weights, Archimedes had got to the same point as Aristotle, but by an entirely different route: instead of deriving his principles from the general laws of motion, he rested his theoretical edifice on a few simple and certain laws of equilibrium, and thus made of the science of equilibria an autonomous science owing nothing to the other branches of physics: he founded statics.

Duhem does not discuss the advantages of this method of proceeding, but the disadvantages, perfect clarity and extreme rigour bought at the expense of generality and fertility: new problems required the invention of completely new independent principles to treat them. Perhaps this was only a temporary loss, to be won back by Archimedes's mechanical successors?[21]

> . . . the certitude and clarity of his principles largely derive from their being drawn, so to speak, from the surface of the phenomena and not dug out from the foundation of things; in a phrase applied less appropriately by Descartes to Galileo, Archimedes "explains very well *that it is* but not *why it is*". So we shall see the most substantial advances in statics come from the teaching of Aristotle rather than the theories of Archimedes.

Unlike Duhemian physics, Aristotelian mechanics did search for explanations, was not autonomous, but was genuinely fruitful. It is possible that at that date this concession was motivated by some lingering sympathy for Scholasticism, but whatever lay behind it, this was the area Duhem was now investigating, and his methodological principles now gave him little guidance.

4. The False Start of the *Études*

Duhem had thus, in my view, worked himself into a historiographical crisis: he had refuted a historiographical programme according to which the Middle Ages was scientifically null, but he had no way of giving a coherent account of the refuting evidence: doing so was liable to put his own methodological principles at risk, and was incompatible with the kind of Machian rational reconstruction that had served him thus far; writing its history was incompatible with maintaining the kind of separation of physics and metaphysics the Machian procedure involved, because it required him to describe the interaction between the two. It required what is known as an externalist historiography, externalist because it explicitly allowed for non-scientific factors, external to science as such. Duhem did not solve his historiographical problems all at once: the required externalist historiography took a while to develop. His initial response to the discoveries of the *Origines de la Statique* was to accumulate more evidence, and do so on a heroic scale: the three volumes of the *Études sur Léonard de Vinci* (1906–13) were to be the result of his efforts, and to do more than anything else to make his reputation as a historian of mediaeval science. Only later came the more considered response of his *To Save the Phenomena* (1908), and *Système du Monde* (1913–59).

The wide influence of the *Études* requires no demonstration. It may well be the most widely-cited of all Duhem's historical works. Directed as it was to the simple accumulation of evidence for mediaeval science, it is fragmentary in nature, consisting largely of articles published independently in journals. Its faults largely derive from that fragmentary nature, and have been widely commented on: its internalist approach leads often to an apparent insensitivity to the context of the 'science' that Duhem found in such abundance, a frequently theological or political context liable to change the meaning of what Duhem found. But fragmentary as it was, and despite its faults, it was complete, and the implications of its main theses were explosive enough ideologically to cause widespread debate. The inevitable result was that critics were liable not to notice when Duhem later revised or refined his judgements. Thus, as will be argued in the next chapter, the *Études* played the same rôle in the reception of Duhem's historical work that Duhem's early 'Physique Expérimentale' played in the reception of his methodological ideas.

It all stemmed from the discovery, recorded in the preface to volume i of the *Origines*, and the source of much of the internal inconsistency of the work, of the non-isolation of

Leonardo da Vinci. Contrary to tradition, Leonardo's work was known about and influential, partly in the work of such plagiarists as Geronimo Cardano—note the practising physicist's characteristic obsession with priorities. Moreover, Leonardo's ideas too had their sources in the writings of the previously despised mediaeval scholastics who, this patriotic Frenchman was happy to discover, were mostly alumni of, or teaching at, the University of Paris. As the art historian Henry Lemonnier[22] was to remark in the course of an appreciative review in 1917, Duhem was not quite without predecessors in these discoveries: his old enemy, the republican scientist and senator P. E. M. Berthelot, had published a note on engines of war in Leonardo's manuscripts,[23] in the course of which, he had this to say:

> The great man's panegyrists have somewhat overplayed his significance when allowing themselves sometimes to attribute to him the discovery of facts and the invention of devices reported in his notebooks. In reality, a large number of these facts and inventions, and the theorems of geometry and perspective he records, as well as enunciations in mechanics and optics were known, before Leonardo . . . it would certainly be a major task to reconstruct the sources Leonardo da Vinci drew on. On certain points this task has been undertaken . . . notably by Mr. G.- B. de Toni . . .

Duhem made his initial aim explicit in the subtitle of the first two volumes: *Ceux qu'il a lus et ceux qui l'ont lu* (Those he read and those who read him). It was to reconstruct, for mechanics at least, Leonardo's sources and his influence, with the emphasis very much on the former. As usual, he published his results as he went along, in all but a few cases as substantially independent articles in the *Bulletin Italien* and *Bulletin Hispanique* of the University of Bordeaux, the whole spread over nine years from 1905 to 1913. Inevitably there is some repetition—even within the first series two articles related to Albert of Saxony—and, equally inevitably over such a long period, there are changes of view. As Henry Lemonnier noted,[24] the *Système* deprives Albert of Saxony of the originality ascribed to him in the second volume of the *Études*. Particularly prominent is the shift of perspective advertised by the change of subtitle in the third volume to *Les précurseurs parisiens de Galilée*, from the rôle of Leonardo in the history of science to the mediaeval sources of Galileo's achievements, from the double task of evaluating Leonardo as the precursor of modern science and finding *his* precursors, to that of finding the precursors of modern science *tout court*. Still, precursors, and scientific ones at that, remain the centre of interest. Duhem might notice that his subjects belonged to this or that (mostly Franciscan) religious order, but he concentrated on their science, and their (perhaps dominant) theological concerns tended to get consigned to the notes. Such a narrow focus is open to the criticism, considered by Duhem himself in relation to Mach, that it omits the intimate relationships between mechanics and philosophy and theology, and the point has often been made since his time. So far, Duhem's work was not only fragmentary and incoherent, but it violated his own canons of sound practice. He may have abandoned rational reconstruction, but the result is still not history.

Duhem's historiographical inadequacy had, however, compensating advantages. He had, it had to be admitted, made substantial discoveries like the theory of impetus discussed above, and a mediaeval semi-mathematics of the variations of intensities of qualities (the so-called doctrine of the latitudes of forms) clearly relevant to the mechanical achievements of the sixteenth and seventeenth centuries. The massive documentation of the sources of Leonardo's mechanics he provided was found convincing, not only by Lemonnier, but by many others since, and dissenters have been forced to discuss Duhem's case if they were to restore (in their eyes anyway) the stature of their Great Man. Similarly, the alleged discovery of precursors for Galileo forced itself on the Galileo sycophants likewise defending the stature of their hero. Much of both controversies can be followed in the abstracts published in the annual volumes of *Raccolta Vinciana*. Duhem's heroes Jean Buridan, Nicole Oresme, François de Meyronnes, have since become the staples of courses[25] in mediaeval science, and it is well known that Galileo's technical term *impeto* had a history going back at least to Buridan in the fourteenth century as had elements of the mathematical apparatus Galileo used to handle the problem of free fall. Moreover, the impoverished character of the history Duhem provided may itself have made it easier to insinuate into scholarly debate Duhem's view of the fourteenth century as the culmination, not the decadence, of Scholasticism, and of the crucial importance of the condemnations, in 1277, by Étienne Tempier, Bishop of Paris, of a long list of Aristotelian theses. The ideological implications of both were already likely to hinder their acceptance: a full Duhemian historical treatment would surely have increased resistance. Both made the growth of science depend on theology and ecclesiastical politics, and both ran counter to an officially sponsored view of the Middle Ages that gave supreme significance to the work of Thomas Aquinas in the third quarter of the preceding thirteenth century. They were never fully accepted but, stripped of a full historical interpretation, they insinuated themselves in a confused way into discussion and have stayed there ever since.

The *Études*, moreover, must have done one other thing: while working on them, Duhem accumulated material, not just of a scientific kind, but philosophical and theological as well, because of the way he was forced to work. With only a single sight of any manuscript he looked at, because everything had to be sent to Bordeaux, he systematically copied everything he might need into notebooks which steadily accumulated. When he went to write a closer approximation to real history he was prepared.

IX

Saving the Phenomena and the System of the World

1. Background and Prehistory of *To Save the Phenomena*

First serialized in the April to September 1908 issues of the *Annales de Philosophie Chrétienne*—the journal whose editorial policy I discussed in the latter part of Chapter III—*To Save the Phenomena* is perhaps the most controversial of all Duhem's works, and the easiest to misinterpret if not read with sufficient care. In it all the various criticisms of his work seem to come together: excessive positivism; neo-Scholasticism; apologetic for the Roman authorities. But it cannot without qualification be labelled both neo-Scholastic and positivist, or both positivist and Catholic apologetic. It matters little that the same people do not offer all of these characterizations: the mere fact that all of them are plausibly offered shows that something is wrong with the way Duhem's work has been read. To remedy this confusion I propose, as elsewhere in this essay, to consider both external and internal factors, both the wider background to the work, and its internal structure. Given my reading of the *Théorie Physique*, it should follow that unless Duhem was grossly inconsistent he cannot here be giving a purely positivist or Scholastic account of events, and given the journal in which it was first published, apologetic for the Roman authorities is unlikely to have commended it for publication. But if my readers will attend to the prehistory of its main theses they should be able to distinguish those that are properly due to Duhem from those he merely took over from the tradition and reworked for his own purposes.

It is famous for two things: its account of the ancient methodological tradition in astronomy that, supposedly, instead of asking whether its hypotheses were true, considered only how effectively they 'saved the phenomena', were compatible with the observations; and its suggestion that in the Galileo affair logic was on the side of Bellarmine and Pope Urban VIII, the ecclesiastics more usually seen as Galileo's blinkered persecutors. The first of these attracts the characterization of 'positivist', and the second that of Catholic apologist. To obtain a clearer view of the work we have to add a third item to our list: the appreciation of ancient astronomy in its own right as a respectable science, and not just as clever rubbish irremediably vitiated by its supposition that the Earth was fixed—as a science, that is, comprising theories of interest independently of the truth of the hypotheses they rested on. It will be clear at the outset that this third aspect is related to the first, for if the truth of the hypotheses in play are of secondary importance, it becomes legitimate to consider the various astronomies of Aristarchus, Apollonius, Hipparchus, and Ptolemy as mathematical systems to be judged by their empirical success, in the same way as modern scientific theories would also be judged. *It is essential for the reader to realize that none of these three theses was Duhem's own.* Though he did indeed play a rôle in elaborating them (mostly before *To Save the Phenomena* was written), all three form the starting points of his argument, not its conclusion. It is unfortunate for a full understanding of his work that its very brilliance obscured the work of his predecessors. Its story in fact parallels that of the *Théorie Physique*, whose brilliance overshadowed the positivist sources of the arguments Duhem here too reworked for his own non-positivist purposes.

It is not as if Duhem hid his sources. His preface duly cited[1] the work of T. H. Martin, Giovanni Schiaparelli, and Paul

Mansion, and the following pages are full of citations to the work of other scholars. The three named, however, are crucial to the story. Martin seems to be a somewhat obscure liberal Catholic with interests in natural theology who contributed to the *Annales* under Bonnetty in the 1870s and 1880s, and had a long track record of research into the history of ancient astronomy. Schiaparelli was the indefatigable searcher after life on other planets who 'discovered' the Martian canals. Paul Mansion was a Belgian Catholic mathematician with interests in the history of mathematics and an association with the Société Scientifique de Bruxelles and its *Revue des Questions Scientifiques*. Along with him one may mention the Jesuit Jules Thirion of Louvain, the editor of that journal. As far as I can tell it was Martin and Schiaparelli who first discovered and explored the saving-the-phenomena tradition, and since neither of them were anywhere near being conservative Catholics, it is safe to presume that defending the authority of the Roman authorities was not their concern, *which was the fate in antiquity of the sun-centred system of Aristarchus of Samos*. Getting on for two thousand years before Copernicus, Aristarchus, whom T. L. Heath was to dub "the Copernicus of antiquity",[2] proposed a sun-centred system to explain the motions of the planets, but that system was soon abandoned in favour of the Earth-centred systems that held the field up to the Renaissance, and its only lasting contribution seems to have been to developing the mathematics of the successful Earth-centred systems of Hipparchus and Ptolemy. The solution Martin and Schiaparelli offered for this puzzling turn of events was that the ancient astronomers were not primarily concerned with the truth of their hypotheses, but with how well they saved the phenomena (technical phrase), how well their theories fitted the often puzzling observations of the motions of the planets.

Mansion and Thirion are another matter. They were Catholics and their concern, in part at least, apologetic. They were also in touch with the harmonizing version of the neo-Scholastic programme associated with the name of Désiré Mercier and discussed above in Chapter II. Being involved in a harmonizing programme meant attending to what non-Catholics were saying, including, in their case at least, what the positivists were saying about scientific method. Thus in 1900, Thirion published in the *Revue des Questions Scientifiques* his three-part essay *Pour l'Astronomie Grecque*[3] in which, making explicit reference to the modern movement in the theory of scientific method, he presented the history of ancient astronomy as that of mathematical systems which ought to attract our admiration, not our disdain. Mansion likewise, in 1891[4] defended Copernicus's publisher, the Lutheran theologian Andreas Osiander, against Kepler's accusations of bad faith. Osiander, it may be recalled, had, in an attempt to ward off criticism, inserted into the published edition of Copernicus's *De Revolutionibus Orbium Coelestium* an anonymous preface of his own[5] which told readers not to worry about the truth of the hypothesis of the motion of the Earth, because truth in these matters was unattainable without divine revelation, and was irrelevant to astronomy anyway. Taken literally, the hypotheses of all astronomical systems were absurd. In the following year, Duhem was to cite[6] that preface as if it was Copernicus's own, and so bring down on himself the wrath of his critic Eugène Vicaire,[7] who cited Mansion's piece against him. Basing himself on Schiaparelli,[8] Mansion had asked why the ancients had preferred Ptolemy to the sun-centred system of Aristarchus, and given a twofold answer: terrestrial geographic co-ordinates and astronomical predictions were easier in an Earth-centred system and:

> because the ancients, who reserved to physics—what we now call cosmology—the search for the causes of the motions of the stars and speculations on the reality of these motions, saw in astronomy nothing but the knowledge of celestial phenomena, and so the choice of hypotheses was to them a matter of extreme indifference.

In support of this claim Mansion cited a passage in Simplicius ascribed to the ancient astronomical writer Geminus, who was summarizing the views of the Stoic Posidonius[9], and claimed that Ptolemy was of the same opinion. He also cited from the commentary of Aquinas on Aristotle's *De Caelo* the claim that while astronomers were not required to use true hypotheses to save the phenomena of the heavens, Artistotle used true suppositions to explain the qualities of their motions. He went on to suggest that though Copernicus sometimes spoke in a contrary sense, in most of his work the talk was about hypotheses and suppositions, in a way that could most easily fit into this tradition. He noted the early tradition of interpreting Copernicus in the manner of Osiander, and cited the parallel modern tradition, as he saw it, of Saint Venant, Jacobi, Kirchhoff, and Poincaré. He seems to have expressed himself in like terms in a work *Sur les Principes fondamentaux de la Géométrie, de la Méchanique et de l'Astronomie* which, though cited by Duhem and indeed printed, never seems to have been published.[10]

By thus suggesting that modern thought on the nature of physical theory was coming back to the ideas of Geminus and Ptolemy, Mansion was offering his 'Saving the Phenomena' tradition as a respectable ancestor for modern ideas on the nature of rational mechanics and mathematical physics. It was too good a gift for Duhem to refuse: in July 1893 he took the hint and replied to Vicaire in these terms:[11]

> Those who resist the above thesis readily claim to be supported by tradition. According to them, all the great thinkers and all the great scientists considered physical theories as trials, as progress towards a metaphysical explanation of things. All of them searched, not to classify the phenomena, but to discover their causes. It was the hope of explaining physical effects that gave them the courage to pursue their researches and their fertility shows us clearly that this hope was no illusion.

Duhem, however, disagreed:

> From the historical point of view nothing is more erroneous than this way of regarding the tradition. With regard to the relations between physics and metaphysics, Aristotle and the peripatetic philosophy accepted a thesis which agrees in essentials with the one we have developed. They scarcely applied it beyond astronomy, at that period the one developed branch of physics, but what they said about the stars is easily extended to other natural phenomena.

He then reproduced Mansion's account, with the addition of some material on Archimedes, before proceeding to discuss the seventeenth century, where a revolution followed "the decadence of Scholasticism":[12]

> Then the ancient barrier separating the study of physical phenomena and their laws from the search for causes is seen to disappear; then we see physical theories taken for metaphysical explanations, and metaphysical systems seeking to establish physical theories deductively.
>
> The illusion that physical theories attain the true causes and the real reason for things penetrates in all directions the writings of Kepler and Galileo; the arguments making up the trial of Galileo would be incomprehensible to anyone who did not see in them the struggle between the physicist wanting his theories to be not only the representation of the phenomena but their explanation as well, and the theologians who maintained the ancient distinction, and did not accept that the physical and mechanical reasonings of Galileo had any force against their cosmology.

We might well ask whether Duhem was in any way interested in maintaining the cosmology of the theologians, or whether he had at that date thought through what he was saying, but let that pass. This passage contains all the elements that were to reappear in *To Save the Phenomena*, but without its confident imperialistic tone, and every one of these elements was to undergo a subtle alteration as Duhem came to realize and respond to the difficulties of the Mansion thesis, for Mansion's it was and not Duhem's.

Written late in 1904, the corresponding passages in the *Théorie Physique*[13] reflected Mansion's extended version of

1899.[14] With a much better documentary base, Mansion was now able, thanks to the work of the Austrian Jesuit ecclesiastical historian Hartmann Grisar,[15] to give much more attention to Galileo. In its essentials, however, his story did not change: Aquinas still represented the whole period between Ptolemy and Copernicus. He made a few changes of substance, adding, with comment, a quotation[16] illustrating Kepler's overconfident astronomical realism, and a citation, taken from Grisar, from a famous letter, to be discussed below, of Cardinal Bellarmine of the Holy Office (which had charge of this and all such cases) to Galileo's ally Paolo Foscarini. Writing at the same time as Thirion, he now also included Duhem in his list of the chief modern writers on scientific method. A *To Save the Phenomena* along these lines would present few problems of interpretation: it would indeed be authoritarian Catholic propaganda, and very sloppy history; it would have been a relatively straightforward application of a positivistic point of view to the historical 'facts'. But the work Duhem finally wrote was altogether more subtle and complex.

2. The Argument of *To Save the Phenomena*

At 140 pages it is a good deal longer, and while still summary and schematic it is altogether richer in historical detail: Aquinas no longer has to do duty for the whole of the Middle Ages, there are 13 pages on Jewish and Islamic cosmology, 19 on the Renaissance before Copernicus, 32 on the period between Copernicus and the Gregorian calendar reform, 29 on the Galileo affair, and a five-page conclusion. On closer inspection it emerges that Duhem has now got into the primary sources and is less dependent on the secondary ones, though there is a change in the character of the latter and more of them to acknowledge: T. H. Martin[17] is added to

Schiaparelli (whose *Origine* replaces the earlier *I Precursori*) and Mansion; the Jew Moritz Steinschneider[18] serves for Ibn al Haitham, and the well-known freethinker Ernest Renan for Averroes;[19] for the Galileo affair the Italian anticlerical deputy Domenico Berti[20] replaces the Jesuit Grisar.

But the changes in Duhem's treatment of the theme are not merely bibliographical: They affect matters of substance as well. I mention two of these, the first of relatively minor, and the second of major importance. In the preface for example, the dogmatic statement that the Scholastic tradition anticipated Duhem's methodological views and not those of his opponents is replaced by the weaker claim that the ancients had debated current methodological questions in different terms.[21] A shift of that sort should warn us that Duhem may have moved away from the narrow perspective of the Mansion thesis and towards a wider approach. The next feature will confirm this suspicion: Mansion's story, and Duhem's in the *Théorie Physique*, considers only the supposed past of a methodological doctrine of superficially positivistic type; *To Save the Phenomena* complicates the story by relating aspects of the scientific and philosophical background to the methodological pronouncements. This second shift of perspective is very easy to miss: questions of methodology dominate the title and much of the treatment, which in any case seems to reflect the superficially instrumentalistic point of view of much of the *Théorie Physique*, but it is fundamental, and the analysis that follows is primarily directed towards bringing it out, for it is essential to the task of assessing the intentions behind the work. *Any interpretation of it depends on a reading of the intentions behind these shifts and additions.*

In the view of Duhem and of many others since, the ancients, concerned with rationalizing the motions of the planets, evolved two different types of theory: on the one hand a philosophical account (derived from the views of Aristotle) based in principle on a doctrine about the nature of the heavenly matter, and in practice on the so called *homocentric* spheres (sharing a common centre with the Earth) of Eudoxus and Calippus; on the other mathematical *epicyclic* systems (with spheres rolling round other spheres) culminating in Ptolemy's *Almagest*. As likely as not both derived through Plato from Pythagorean speculations. But these two theoretical approaches were impossible to reconcile, and the resulting problem was dealt with in various ways including, in extreme cases, an implausible division of labour theory according to which the philosophical 'physicist' told the truth about the heavens, while the 'mathematician' created geometrical systems whose sole purpose was to 'save the phenomena' observed in the heavens, using only smooth, uniform circular motions. This distinction, which may be compared with Scholastic and Pascalian ways of classifying the sciences discussed in Chapter V, did not satisfy everyone: the Averroist tradition, for example, deriving from the Spanish Muslim Ibn Rushd (Averroes), tried to insist that the only good mathematical astronomy was one based on true hypotheses, that is one that only used smooth, uniform, circular motions whose centres coincided with the centre of the world, but all attempts to construct a *homocentric* astronomy satisfying those conditions failed the test of even the crude observations then available.

As these arguments developed there appeared the two impossibilities that interested Duhem: that of deducing a mathematical system from a metaphysics (which he took to be equivalent to the ancient 'physics') and that of demonstrating from observations the truth of a mathematical system. Here Duhem, never afraid of awarding marks to the people he studied, inevitably awarded good ones to those who insisted on these points. This is what leads him in his 'Conclusion' to make his classic statement[22] that "Logic was on the side of Bellarmine . . . ", rather than on that of Kepler and Galileo. All these judgements, however, are complicated, in a way many of his readers have missed, by cosmological considerations added to the story Duhem took over from his predecessors: adding the distinction between earthly and heavenly physics greatly changes the distribution of marks.

3. Heaven and Earth

This second distinction is hardly controversial. Aristotle held in effect that physics changed at the orbit of the moon: below was generation and corruption, change and decay, forced and violent motion, motion in straight lines, and all sorts of other 'irregularities'; above was the unchanging eternal 'quintessence' where the only possible motion was in unvarying uniform circles round the centre of the world, for what else but such eternally unvarying perfection could suit the abode of the gods (the idea that the stars were 'visible gods' was to provide one kind of backing for ever-popular astrology). It is surely a commonplace that the cosmology that gave meaning to the earth-heaven distinction was replaced during the seventeenth century by others, culminating in that of Newton, in which it had no place. Less obvious, perhaps, is that it was pagan: for the orthodox Christian (and also for the Jew or Muslim) who knows that God is a spirit who is everywhere and nowhere at the same time, the visible skies are part of God's creation, distinct from Him, and conventional imagery should not disguise this. Both of these aspects of the ancient distinction between the physics of earth and heaven concerned Duhem. The first is specifically mentioned in his 'Conclusion', while the second is discussed in the 1911 letter to Bulliot discussed above. It too leads to the award of marks.

Duhem's concern with this second distinction can be seen in his treatment of the Platonist Proclus and the Jew Maimonides. Duhem reads the former as saying:[23]

> . . . astronomy . . . does not grasp the essence of celestial things: it gives only an image of it, and that very image is not exact, only approximate; . . . the geometrical artifices serving us for hypotheses to save the apparent motions of the stars are neither true nor plausible. They are pure conceptions that could not be realized without formulating absurdities.

Apart from natural classification, these may seem irreproachably Duhemian sentiments, but the reason Duhem reads Proclus as giving for them are not:

> . . . these features of Astronomy should not surprise us. They only indicate that human knowledge is limited and relative, that human science could not emulate divine science.

a point that can be compared with Duhem's translation of a passage from Proclus's *Commentary on the Timaeus*:[24]

> When sublunary things are at issue, because of the instability of the matter forming them we are content to take what occurs in the majority of cases, but when we want to know celestial things, we use feeling and appeal to a crowd of artifices far removed from any plausibility. Hence, as regards each one of these, we must be content with the nearly so, we who are lodged, as is said, in the lowest place in the universe. The gods, for sure, have a more certain judgement.

Thus the difficulty in the way of getting the true astronomical hypotheses is grounded not in considerations of logic relating to the structure of mathematical theory, but in the subject-matter itself: the heavens, the abode of the gods, are far above the comprehension of finite men. Duhem has already quoted Ptolemy as saying:[25]

> It is inappropriate . . . to compare human things with the divine; we must not found our confidence concerning objects placed so high by relying on examples taken from what is most different from them. Indeed is there anything differing more from unchangeable beings than continually changing beings? From beings subject to the constraint of the

> entire universe than beings freed even from the constraint they themselves exert?

He now presents Maimonides as taking a very similar line:[26]

> The idea dominating all Maimonides's astronomical discussions, the new idea within Semitic Aristotelianism, and which in that environment astonishes by its prudently sceptical aspects, is the idea already pointed to by Ptolemy and developed by Proclus: the knowledge of celestial things, of their essence and of their true nature, is beyond man's powers: sublunary things only are accessible to our weak reason.

and quotes from the *Guide of the Perplexed*:[27]

> Know that if a mere mathematician reads and understands the astronomical subjects under discussion, he can believe that at issue is a decisive proof demonstrating that that is how the form and motion of the spheres is. But it is not so, and that is not what astronomical science seeks . . . the aim of this science is to suppose a system by which the motion of the star could be uniform, circular, never accelerated, nor retarded, nor changed in direction, and whose result is in agreement with what is seen. . . .

Again this seems thoroughly Duhemian, but the explanation for this state of affairs is not:[28]

> . . . everything Aristotle said about sublunary things has a logical coherence: these are things whose cause is known and are deduced from each other, and the place in them held by the wisdom and providence of Nature is evident and manifest. As for everything in Heaven, man knows nothing of it but this little of mathematical theories, and you see what that amounts to. To use a poetic locution, I will say: "The heavens belong to the Eternal, but he has given earth to the sons of Adam" (Psalm 115, v.16). God alone that is knows perfectly the true nature of Heaven, its substance, its form, its motions, and their causes . . . It is impossible for us to have the necessary elements to reason about Heaven, which is far from us and too high in its position and rank.

The reader will perhaps recall at this point Osiander's classic preface, quoted by Duhem later in his work, where the situation is considered in which different hypotheses are offered to account for one and the same motion. In such cases, said Osiander, the astronomer will prefer the hypothesis that is easiest to grasp, while the philosopher will rather seek the semblance of the truth. Neither, though, will be able to conceive or formulate any certainly true hypothesis *without divine revelation*.[29]

The consequences of this cosmological factor are far-reaching: the view of the nature and significance of astronomical hypotheses Duhem approves of was embedded in, and supported by, not logical arguments, but arguments derived from a natural philosophy and (Duhem was later to emphasize) a theology he had no use for at all. The point is not just a later discovery of Duhem or of critics like Clavelin[30] half a century later, but it was already explicit in the work he wrote. It was, I believe, the key to his argument, both in the body of his text and his final conclusion, where he shows his concern to establish what was right about the views of Copernicus, Kepler, and Galileo, to set against what was wrong:[31]

> The Copernicans stuck stubbornly to an illogical realism, when everything was pushing them to give up that error, when by attributing to astronomical hypotheses the true value so many authoritative men had established, it would have been easy for them to avoid at the same time the disputes of the philosophers and the censures of the theologians. Puzzling conduct, calling for explanation! Now can it be explained in any other way than by the attraction of some great truth, a truth too vaguely perceived by the Copernicans for them to formulate it cleanly, to separate it from the erroneous affirmations behind which it had hid, but a truth felt with so much vivacity that neither the precepts of logic nor the counsels of interest could weaken its invisible attraction?

The truth at issue was the physical unity of earth and heaven:[32]

> While Kepler multiplied his attempts at accounting for the motions of the

> stars with the help of the properties of streams of water and magnets, while Galileo sought to harmonize the flight of projectiles with the motion of the Earth or to draw from the latter motion the explanation of the tides, they both believed they were proving that the Copernican hypotheses were founded in the nature of things, but the truth they were gradually introducing into Science was that a single Dynamics ought, in a single ensemble of mathematical formulae, represent the motions of the stars, the oscillations of the Ocean, the fall of heavy bodies. They thought they were updating Aristotle; they were preparing for Newton.

Duhem has perhaps been working too quickly, for his thought is uncharacteristically vague. Perhaps his argument can be rephrased thus: what was, by Duhem's criteria methodologically right, was understood as a consequence of what was cosmologically wrong. Those who were revising the cosmology, or the associated mathematics, could hardly avoid rejecting what they saw as the epistemological consequences of what they were busy dismantling, for from their perspective they could hardly distinguish the two. The one had to go with the other. As Maurice Clavelin put it in 1964:[33]

> To accept the hypothetical equivalence of hypotheses put to him by Bellarmine and the philosophers could . . . for Galileo mean only, in the context of the time, one thing, viz., the right of traditional philosophy to continue to lead and guide scientific analysis properly so-called.

4. Understanding Duhem's Historical Revisionism

Noteworthy about Duhem's analysis is the extent to which it reflects the conventional wisdom of its time. Tradition had it that the ruin of Aristotle was the necessary prerequisite for the arrival of modern science, and Duhem agrees. To Koestler,[34] the Galileo affair was an unhappy accident, but not to Duhem. For him, given the confusion of ideas attending the birth of the new ideas, preparing for Newton was in practice impossible without insisting on the truth of Copernicanism. Equally, the theologians of the Holy Office were within their rights in examining in their terms the two propositions presented to them:[35]

> They asked if these two propositions showed the two marks that both Copernicans and Ptolemaists were mutually agreed in requiring of every acceptable astronomical hypothesis: were they compatible with sound physics; were they reconcilable with divinely inspired Scripture? . . . the two censured propositions showed neither the one nor the other of the two features every acceptable astronomical hypothesis had to possess. Hence they had to be rejected altogether, not even used with the sole aid of *saving the phenomena*. So the Holy Office forbade Galileo from teaching the doctrine of Copernicus in any form.

On which Duhem comments:

> The condemnation brought by the Holy Office was the consequence of the collision of two realisms.

The realisms being that of Aristotle and Averroes (not that of Aquinas, explicitly anyway) on the one hand, and that of Copernicus and Galileo on the other, both sides insisting irrelevantly on metaphysical criteria for scientific truth.

Duhem has thus turned the tradition he inherited, as represented in particular by his friend Mansion, inside out. It is no longer a strategy of vindicating the authority of the Holy Office, nor a means of lending the respectability of antiquity to his own methodological views. Mansion's result is no longer a result but a problem for Duhem the scientist, and the solution to that problem, which hardly seems to favour Galileo's ecclesiastical opponents, has a double-edged flavour that seems to have escaped his critics altogether. Was Duhem being merely naive, or over-subtle, when he suggested that for the theologians of the Holy Office, sound physics was that of Aristotle and Averroes? By 1908 he must have known that

Averroism was not a nice thing for theologians to be accused of. Granted, it was in theology that Averroism was dangerous, not in natural philosophy, but how separable were these two areas? Could it not be that Duhem was sending out signals to those of his readers in the know? The theologians of the Holy Office were in any case Scholastics, followers of the philosophy that the Roman authorities in Duhem's day were doing so much to further. Whatever else Duhem was doing, he cannot have been advertising any commitment to that philosophy.

The signals have been pretty invisible to those of Duhem's readers not involved in his milieu and in the conflicts of his time. To appreciate them, it was probably essential to be familiar with the historiographical tradition he worked in, to know Martin and Mansion, to know Grisar and the attempts made by him and other Catholic apologists to use Scholastic arguments to rescue the reputation of the Roman authorities from the shadow of the Galileo affair. It was essential to observe closely the moves Duhem made on the board these and like predecessors provided him to play on. But Duhem's work replaced theirs in the literature: later scholars did not read them, but on finding their theses in Duhem did not follow his references back to their sources. So when they were surprised by these logical theses in Duhem that were not his prime concern, they did not ask with sufficient care what his purpose was. The message is clear: it is dangerous to use or even criticize an important secondary source without first doing just that, without as recommended by R. G. Collingwood,[36] doing the history of your historical problem.

It is doing the Collingwoodian history of the history that reveals the extent to which Duhem has been misread: as a historical exercise in applied methodology, with the aim of arriving at a methodological judgement on events like the Galileo affair, when it is at least as concerned with cosmology as methodology: and as an application of an instrumentalist methodology, in which nothing was allowed of the scientist beyond what was licensed by observation, when this was far from Duhem's position. As already argued at length in Chapter VI, for Duhem "logic is not the sole guide of our judgements": getting your logic wrong is not necessarily the end of the matter. Galileo may get low marks on that count, while still getting high marks for his contributions to the advancement of modern physics, for what, as Duhem sees it, *is ultimately the same reason*, the two issues were the two sides of the same coin. For Duhem as for Pascal, "We have an idea of the truth invincible by any pyrrhonism".[37] The goal of physics is a natural classification unendingly approached indirectly through the successive reclassifications of new physical theories. Here *finesse*, the intuitive faculty, is the judge, and in the reclassification of the world involved in the abolition of the earth-heaven distinction, it surely had to have a prominent place.

All that is so easy to miss. My analysis of *To Save the Phenomena* has concentrated on those parts of the work that show most clearly Duhem's revisionism relative to earlier, often Scholastic, discussions of its themes, and on the rôle it gives cosmology beside methodology. But as remarked above methodology does indeed dominate the bulk of it, and with that comes a superficially instrumentalist flavour that seems somewhat cruder than that of parts of the *Théorie Physique*. From that perspective it is easy to assume that this is the point of the work, the sole basis on which Duhem is awarding his marks, and to assume with most of his other readers, as I too once did, that the cosmological elements obviously involved in the views of Ptolemy, Proclus, and Maimonides are inconsistent with Duhem's apparent thesis. Even those familiar with the earlier writings could easily note the wording of the title, and the extent to which it extended the methodological arguments of Martin and Mansion, and so conclude that that was all it was about, despite the obvious prominence of cosmological considerations in Duhem's conclusion. Nevertheless, my analysis has,

I think, shown that for Duhem, as for Imre Lakatos,[38] we know what great science is and that Copernicus, Kepler, and Galileo practised it. *To Save the Phenomena* chalks up one more victory for *bon sens* and *finesse* over *géométrie*.

5. The Theme of the *Système du Monde*

Though advertised in its subtitle as an *Essai sur la notion de théorie physique de Platon à Galilée*, *To Save the Phenomena* had turned out to be as much about the history of cosmology as about methodology. It emerged that the idea of a physical theory depended on the corresponding cosmology and could not be understood without it. If then, this *Essai*, this trial, was to become the basis of the full history it surely calls for, that history is going to be one of cosmology, in which methodological questions concerning the relation of physics and metaphysics will play only a relatively subordinate rôle. Cosmology is indeed the explicit theme of Duhem's major historical work, *Le Système du Monde, Histoire des doctrines cosmologiques de Platon à Copernic*. Intimately related to *To Save the Phenomena*, texts discussed in the latter are usually discussed at the appropriate place in the former in almost identical terms, though often in a new context that is liable to put them in a new light.

Needless to say, however, closer inspection reveals once again some major shifts of interest, and the work that emerges is rather different in character from that promised in the title. Some of these shifts were probably concealed from many readers by the unfortunate publication history of the work, interrupted as it was by Duhem's death in 1916 and only completed in the 1950s after much lobbying by *inter alios* George Sarton and Marie Tannery,[39] but even volumes i–v, available from 1917, are enough to show up most of the significant shifts.

In the first place there is the extension of 'cosmology' to include terrestrial as well as celestial physics, which ensured that all the material on pre-Renaissance mechanics in the *Origines*, the *Études* and elsewhere would naturally find a place in it, and also enabled Duhem to supply some of the context for the speculations on the motion of the earth by impetus theorists like Nicole Oresme and François de Meyronnes.[40] Already a main reason for the scale of the work, this extension is easily understandable if Duhem's main concern is with the prehistory of the *end* of the earth-heaven distinction and of the peripatetic physics it was part of, for in the coming of the new physics both would have their part to play. However, understandable as it is, a work covering such a wide range of topics is hardly ever going to be read right through, particularly in view of the span of time it covers. It is more likely to be looked up for guidance or information on particular topics often only treated incidentally as part of a larger argument.

In the second place, there is the inclusion of philosophy as well as physics. Though the need to deal with contemporary neo-Scholasticism may also have been a factor, this was in part at least a consequence of the prominence Duhem, as already in the *Études*, gave the 1277 condemnations in Paris and Oxford of certain Aristotelian theses of a basically philosophical kind.[41] Insofar as they were relevant to Duhem's story, it would have to be through the interaction of philosophy and physics. Duhem therefore gave expression to his own belief, voiced in the Mach review discussed above, that the history of pre-modern physics was intimately involved with philosophy. The treatment he gave was hardly superficial. Parts iii ('La Crue de l'Aristotélisme'—the flood-tide of Aristotelianism) and iv ('Le Réflux de l'Aristotélisme—the ebb of Aristotelianism) represent about two and a half published volumes in all, and are on little else, while philosophy obtrudes persistently throughout the rest of the work.

In the third place there is theology. It too was required by the comments on Mach referred to above, but in any case it was now essential to his overall thesis. His 1911 letter to Bulliot had declared that:[42]

> From its beginning Greek science was altogether steeped in theology—but the theology was pagan: theology taught that the heavens and the stars were gods; it taught that they could have no other motion than the circular and uniform motion that is perfect; it cursed the blasphemer who dared attribute a motion to the Earth, the sacred focus of the divine.

He was also to declare in the crucial opening paragraph of Part v:[43]

> From the beginning of the fourteenth century, the grandiose edifice of peripatetic physics was condemned to destruction: the Christian faith had undermined all its essential principles; observational science, or at least the only observational science then developed a little, Astronomy, had rejected their consequences; the ancient monument was going to disappear; modern science would replace it.

It is unfortunate indeed that this sentence summarizing all the essential features of Duhem's final position was not available in print until the 1950s. Its third clause refers back to the preface and conclusion of *To Save the Phenomena*, and the second is blatantly subversive of contemporary neo-Scholastic orthodoxy. It makes explicit, what must have long become obvious, that notwithstanding his advertised refusal, in relation to Aquinas, to pass judgement "in the properly theological area",[44] Duhem was up to his neck in theology, and only wished as a Catholic layman to appear not to be trespassing too obviously on the sacristy.

Duhem was also engaged on an enterprise whose completion was always in doubt. Discussing his last days, his friend Édouard Jordan reports:[45]

> If apart from paternal affection, not to be compared with anything else, one thing could have held him to life, it was the *Système du Monde*. When he began, I asked him one day, certainly not with foreknowledge of his premature end, but conscious of the vast scale of the enterprise, if he sometimes feared he might not see its end. "I don't think of that", he said to me. "If God judges this work useful, He will give me the time to finish it. Otherwise, what does it matter?" Despite that resignation, he would have been particularly hurt to leave his monument unfinished.

He certainly worked at enormous if not excessive speed: Starting in 1912–13, he published four volumes in his lifetime and a fifth was in press at his death, while he left behind five more publishable volumes, giving ten out of the intended twelve. He did not finish, and did not live to write the summary he told Jordan he was planning:[46]

> When I have finished my *Système du Monde*, I will shut myself up during the holidays at Cabrespine, and in three hundred pages without scholarly apparatus I will extract from it the essential conclusions.

Nonetheless, what he did achieve, if read with care, is more than enough to tell us his essential theses, though readers may well wonder how Duhem would have handled the revival of Aristotelianism in the fifteenth, sixteenth, and seventeenth centuries. Read sensitively, the *Système* may even yet have something to teach historians of science.

6. The Character of the *Système*

To write ten volumes of a work on that scale in under five years requires a lot of single-minded dedication, even by Duhem's standards, and he had that in abundance. He needed it, because the work was directed to no obvious ready-made intellectual constituency, except possibly the Blondel circle.[47] There was a constituency willing to believe that there was interesting science done in the Middle Ages: it even included Duhem's old enemy Marcellin Berthelot, who would not have been willing to concede that the Church deserved any credit for this fact. There was even a constituency willing to give the Church credit for Mediaeval science, but it consisted of neo-Scholastics not at all ready to swallow Duhem's explanation, that it was due to the Church's rejection of Aristotle. It was central to his position that Aristotelian natural philosophy was not only incompatible with the growth of modern science, but altogether irreconcilable with the Christian faith, and the aim of the *Système* was to teach this lesson. The lesson is taught in a number of ways, both explicit and implicit: by a hostile 100-page analysis of Aquinas; by the prominence already referred to given the condemnations in 1277 of a whole collection of Aristotelian theses including, Duhem was careful to note,[48] some important to Aquinas; indirectly by the importance he assigned to the nominalist schools of fourteenth-century Paris, the period conventionally presented as the decadence of Scholasticism.[49]

Duhem's treatment of Aquinas can be usefully contrasted with that of Pope Leo XIII in the Encyclical *Aeterni Patris*, which asserts:[50]

> Philosophy has no part which he did not finely touch at once and thoroughly: on the laws of reasoning, on man and other sensible things, on human actions and their principles, he reasoned in such a manner that in him there is wanting neither a full array of questions nor an apt disposal of the various parts, nor the best method of proceeding, nor soundness of principles or strength of argument, nor cleanness and elegance of style nor a facility of explaining what is abstruse.

The final pages of Duhem's chapter on Aquinas take a very different view which can be left to speak for itself:[51]

> The huge composition elaborated by Thomas Aquinas presents itself . . . to us as a patchwork in which are juxtaposed, precisely

recognizable and mutually distinct, a multitude of pieces borrowed from all the philosophers of Greek paganism, patristic Christianity, Islam, and Judaism.

Hence Thomism is not a philosophic doctrine: it is an aspiration and a tendency. it is not a synthesis but a desire for synthesis.

Like . . . a child trying to put together the separated pieces of a jigsaw puzzle, Thomas Aquinas juxtaposes the fragments he separates from Aristotelianism and every neo-Platonism, convinced that these pieces, so varied in shape and colour, will finish up by reproducing a harmonious picture, a philosophic image of Catholic dogma.

His desire for synthesis is so great that it blinds in him the discernment of his critical sense. It never occurs to his mind that however they are cut up and distributed, the doctrines of Aristotle, of the *Liber de Causis,* of Avicenna will never end up agreeing with each other, that they are radically heterogeneous and incompatible, and that above all they are incompatible with the Christian faith.

I know of no sustained reply to this chapter: Gilson, for example, ignores it altogether. It could be said that this kind of criticism misses the point, that Aquinas transforms all he touches, which is just what Duhem expressly denies, though in later life his friend Blondel was to come round to a less hostile appreciation of Aquinas, and his Bordeaux colleague Dufourcq thought that here Duhem was going a bit far.[52] It is not my purpose here, however, to judge between the Duhemian and neo-Scholastic interpretations of Aquinas, but to establish just what Duhem's position actually was, and of that there can be no doubt. There can hardly be a more serious criticism of a philosopher than that he lacks critical sense. Serious in any academic discipline, it is doubly so in philosophy where it means that the work concerned is altogether worthless, useless for anything, useless above all for further work and teaching. There is no way that the writer of the above passage can be a neo-Thomist Scholastic of any kind. Not neutral history, this is present-day polemic, with the clear and obvious aim of destroying Thomist Scholasticism root and branch, for the philosophy in which analysis discovers such vices is good for *nothing*.

I have already referred to Duhem's treatment of the condemnations of 1277 in the *Études*. In the *Système* it was further grist to Duhem's anti-Thomist mill. The irreconcilability of Aristotelianism with the Christian faith was not just his own private opinion, but had been authoritatively affirmed by the Church when the problem first arose. It was just unfortunate that that same Church was just then condemning some of his friends for affirming what Bishop Tempier of Paris had also affirmed. Duhem was very well aware that his thesis was controversial: Mandonnet's opinion that the condemnations were not particularly important is cited at second hand in a letter to him from Albert Dufourcq.[53] Duhem, however, did regard them as important, asserting that what was thereby condemned were the bases of the entire system, and he particularly noted the additional theses condemned at Oxford the following year, as well as the theses that particularly affected Aquinas. His analysis and the conclusions he drew from it have remained controversial ever since, though they may not have always been accurately understood: according to him, the condemnations prevented the formation of an Aristotelian orthodoxy standing in the way of the formation of modern science. This, he said, made modern science *possible*, but did not entail that modern science *would* come about.

The crown of the work is the prominence Duhem gave to the (mainly Franciscan) so-called nominalists of fourteenth-century Paris. They had already been the subject of Volume iii of the *Études*, but in the *Système* as we now have it, they occupy three volumes, Volumes vii-ix, in which he ploughed the field he had been working since he first discovered the theory of *impetus* in 1904.[54] This material served more purposes than one: the prominence in his story of the theologians and philosophers of the Sorbonne was undoubtedly congenial to the patriot who as a teenager in the years of the Franco-Prussian war belonged to the generation that could think only of the *revanche* for which the opportunity had now come with the advent of the First World War. Like many others, Duhem saw himself as contributing intellectually to the recovery of France from German cultural domination. We need not doubt that his special affection for and loyalty to Paris in particular was also a factor.

But the patriotic message was in my opinion marginal beside the religious. Fourteenth-century Paris created philosophies, usually referred to as nominalist, that followed up the condemnations of 1277 by ruling out dogmatic Aristotelianism in advance, doing so in a way that may have been decisive for all later Western intellectual history. As viewed by Aristotelian natural philosophers, the world consisted of *substances*—man, animal, plant, mineral, rock, earth, wood, angel, God, all of whose properties could be logically deduced from the particular nature, form, or essence of each one, and this essence, in the Aristotelian view of what science consisted of, was accessible through an *essential definition*. It was a consequence of this metaphysics that the properties and behaviour of these substances was governed by a natural necessity. But that, however, ran counter to most monotheistic theologies, which have always laid great emphasis on God's will, and on His power to do, like an absolute Monarch, just what He liked with His Creation. Hence, as the future Pope Urban VIII, Maffeo Barberini, was long afterwards to reiterate in a celebrated audience with Galileo,[55] philosophical reason was forbidden to limit the power of God beyond the restrictions imposed by the law of non-contradiction. Technically known as *voluntarism*, this attitude was to be a constant feature of much theology and philosophy down to the end of the seventeenth century, and debates about its rôle in the scientific revolution continue.[56]

The theologians could condemn the consequence, but it was better to avoid the principles it followed from, and the method chosen by the nominalists, of whom the most famous was

William of Ockham, was a philosophy that ruled out *a priori* the metaphysics of substances: it was denied that these general substances were anything real in the world, and asserted that they were only *names* for arbitrary collections of individual things. There just was nothing from which to derive these pretended natural necessities. Entities, William of Ockham may or may not have said, were not to be multiplied beyond necessity. He certainly doubted the necessity of the essentially defined Scholastic substances.

It might be though that this heady theologico-philosophical brew would stop all rational inquiry altogether, but it need not, as is evident from its prominence in the thought of such figures as Descartes and Newton, and Duhem claimed that it set fourteenth-century minds free to speculate and to think the previously unthinkable. The doctors of the Sorbonne, whose Catholic orthodoxy, Duhem reminded Bulliot, was famed throughout the entire world, felt free to speculate on the plurality of worlds and the motion of the earth and to develop, as remarked above, a theory of *impetus* in mechanics against the prohibitions of Aristotelian natural philosophy. All this took place in a period that Thomist neo-Scholasticism had written off as a period of decadence, and Dom David Knowles[57] was to deplore as the "harvest of nominalism," in which philosophy ceased to be regarded as a source of definite knowledge.

7. Understanding the *Système*

The overall story has long since become familiar to historians[58] of science. It has remained controversial, and like any other major historical synthesis is eminently criticizable, particularly in view of the polemical intent that is the source of its interest, and the speed with which it was composed. The errors that Otto Neugebauer[59] was able to detect are doubtless only a few among many. Discussing errors of detail, however, is a task that can safely be left to the professionals. My task here has been to put these into context by enabling readers to appreciate and assess Duhem's central historical concerns. One or two final points may assist me towards that aim.

Firstly, just as Duhem did not ascribe any scientific creativity to Bellarmine and Barberini in *To Save the Phenomena*, so likewise he did not ascribe any such creativity to Kilwardby and Tempier, the authors of the condemnations of 1277, or indeed to William of Ockham. They were rather credited with something different: by destroying Aristotelian orthodoxy, making room for the growth of the modern science they themselves did nothing positive to create.[60] Duhem was thus not here giving them the kind of credit his critics may have been misled by the *Études* into thinking he did. As he put it in his 1911 letter to Bulliot:[61]

> If these theological doctrines supplied a few provisionally useful postulates to the science of nature, if they guided its first steps, they soon became what leading strings become for a child, fetters. If the human mind had not broken these fetters, it would not have been able to get beyond Aristotle in physics, or Ptolemy in astronomy. But what broke these fetters? Christianity.

Secondly, what emerges is the thorough conservatism of Duhem's position. He knows that present-day physics is superior to ancient, and is prepared to use that knowledge in his historical argument, talking about obstacles to progress like any old-style scientist historian. He may not think *all* theology to be an obstacle to the progress of science, but he is quite sure that some types of theology are that, and that discerning these obstacles is his legitimate concern. Modern historiographical orthodoxy is that it is the historian's business to understand the past in its own terms without any *arrière pensée* towards what science was later to become, but Duhem always keeps in his mind his understanding of, and concerns, in the present.

Thirdly, Duhem's work is not at all a breakthrough towards a more favourable evaluation of Aristotle. It is rather a sustained attempt at rescuing, in the interests of modern science and his own understanding of Christianity, the Middle Ages and the Church from an excessively Aristotelian perspective. Needless to say, and not for the first time in his life, it put him in a distinct minority. As he put it in a letter to Blondel, at the height of the crisis caused by the condemnation of the *Annales de Philosophie Chrétienne*:[62]

> . . . a definite impression, which increases in proportion as I plough the history of Scholasticism, is that through ignorance or prejudice our neo-Thomists offer us a false Aristotle, a false St. Thomas, and a false Scholasticism, and they understand absolutely nothing of this great intellectual movement of the Middle Ages, which they boast about to us, and it is admirable indeed, but has no resemblance to what they tell us.

Fourthly, the prime focus of this work is the interactions between physics, philosophy, and theology, not the physics as such. It is thus, to use the terminology which became standard for such discussions in the 1960s and 1970s, externalist. The history of cosmology is explained in terms of the history of philosophy and theology, subjects which may sometimes claim to be regarded as sciences, but are not identical with cosmology, and are equally not identical with the mechanics that also plays, as noted above, an important rôle in the story. All that the *Système* has in common with internal history of science, or with the critical history written earlier in his career and discussed above in Chapter VII, is its intellectualist orientation, towards the various types of theory advanced and their problems, whether perceived at the time or by their historian Duhem: there is little sign here of the economic and social factors that have become the staple of modern externalist historiography. But the identification, apparently going back to A.R. Hall's paper 'Merton Revisited', of 'internal' with 'intellectual' is surely a mistake: external intellectual history is as possible a genre as the internal sociological history that might be derived from T.S. Kuhn's *Structure of Scientific Revolutions.*

Fifthly, in common with much external history even where as here the details of the science are taken direct from the primary sources, it will not necessarily be reliable on the technical details of the science under discussion, particularly if we remember the speed with which it was written. Despite his career and professional self-evaluation as a mathematical physicist, Duhem's main concern is no longer with the details of the mathematics but with its philosophical and theological relations, and it is here that his work stands or falls. Even if his work may well be found useful in many other ways, the user or critic has to remember that Duhem is offering a thesis about the origins of modern science as an enterprise and of its main features, and it is as such that it has to be judged.

X

Conclusion: The Reception of Duhem's Work

1. Internal Obstacles to Understanding Duhem

The task I set myself in this study was to understand in its intellectual, religious, and political context the thought of the historical figure Pierre Duhem, with particular emphasis on his philosophical and historical writings. I started from the observation that in the eyes of many, his real stature was problematic, a brilliant maverick who continually got things frustratingly wrong: producing brilliant arguments against atomic explanations in physics and chemistry, a muddled instrumentalism in the philosophy of science, and a voluminous collection of misreadings of mediaeval Scholastics. My study presents on the contrary a brilliant man who was indeed something of a maverick, but a much more subtle one than his opponents have allowed, a man whose work opened up issues that are still not closed. It is my central claim that neither the man's writings nor his context have been considered carefully enough by his critics. But just as there are historical reasons for Duhem taking the approaches he did, there are historical reasons for this lack of careful consideration. These reasons have to do with the historical situation his readers found themselves in, and with the unfortunate consequences of Duhem's own method of working. It is now time to examine these historical reasons, to broach the problem of the reception of Duhem's work. I begin by discussing those factors in its reception that seem traceable to Duhem himself and his method of working, before proceeding to discuss the wider historical factors in the reception, first of the philosophical work, and then of the historical.

On several occasions I have referred to the misunderstandings that arose in the minds of Duhem's readers. Some of these misunderstandings, I have already hinted and will argue further below, are ascribable to the various ideological, philosophical and religious presuppositions his readers brought to his work, others to accidents of publication and the like. But some must be blamed firmly on Duhem himself: there had to be a price paid for the facility in writing that enabled him to get by with single drafts only, without notes,[1] and an even bigger price for the sheer size of his output produced under the atrocious conditions of near-isolation in Bordeaux, with only a single chance to look at any manuscript he wanted to consult. It was all done on the run with no chance to reflect, and so he let pass changes of view that damaged the integrity of his work, changes of view sometimes both exaggerated and partially concealed by Duhem's publication habits.

I have remarked more than once on Duhem's habit of publishing works before they were complete. Manifestly the case with the *Origines*, I have found evidence of it also in the *Théorie Physique*, with its wholesale reprinting of large chunks of material over ten years old. There can be little argument that that was how the *Études* was written, and a detailed analysis of the *Système* would almost certainly reveal evidence of similar methods of composition, though here the scale of the work probably made it inevitable. How far similar things happened in the composition of Duhem's scientific work is hard to say, and would need a great deal of analysis to establish, but in the philosophical and historical work it seems to have been his normal practice.

Though it may reflect a proper sense of priorities in one who regarded himself first and foremost as a physicist, it is not difficult to see the consequences of such practices: at worst unperceived changes of view in the course of composition, works started before he knew where his argument was going to lead, changes in overall attitudes in the course of his career, writing liable to mislead the inattentive reader into serious misreadings.

For example, following Duhem's own suggestion, I have sought in this work for an Ariadne's thread, a line leading through Duhem's work that would reveal the point of each of its parts in relation to the whole. But, significantly, the thread I have identified is not the one Duhem pointed to in his submission to the Académie des Sciences in 1913: there he presented all his philosophical and historical work as extensions of his interests in physics, while on the contrary I have emphasized religious, cultural, and political considerations. This change of perspective is partly the result of concentrating on Duhem's philosophical and historical work, largely excluding from my purview the physics that was and remained the centre of Duhem's academic life, but the problem goes deeper. If the aim of the *Théorie Physique* is the Pascalian one of establishing the superiority and primacy of the *coeur* over *géométrie*, then it is not easy to find a close connection between it and the straightforward apology for the Energetic programme Duhem encouraged readers like Lowinger to see in it. Similarly, if the aim of Duhem's researches into mediaeval science is at least in part the religious one of undermining contemporary neo-Scholasticism and refuting contemporary anti-clerical propaganda, then it ceases to be a straightforward test of the philosophy against the facts of history. Duhem's failure to recognize such shifts is worrying (I take it that I have established their existence): unrecognized shifts of this kind can have immensely damaging effects, because they may make readers less likely to recognize them. In Duhem's case they *had* to be recognized if his work was to be understood, for they are pervasive.

In the course of this essay I have pointed out a number of

examples of such shifts. Lemonnier's example of Albert of Saxony is one case: after initially seeing him as an original contributor to mediaeval mechanics Duhem later saw him as a repeater of the ideas of others. Another is the change of focus in *To Save the Phenomena* from methodology to cosmology. Yet another is the shift in the *Système du Monde* from cosmology to the overall relations of physics with philosophy and theology. These shifts affect individual works, but there is one that may have caused more trouble than any other: the increasing emphasis on the Pascalian methodology of *bon sens* as Duhem's career progressed.

The prime focus of this study has been Duhem's later career, and its prime interpretative key the *Science Allemande* of 1915, the work that spells out Duhem's Pascalian orientation and makes clear just how central Pascal was to the thinking of a man for whom common sense was the ultimate foundation of all knowledge. All the elements there spelled out have been easy to identify in the *Théorie Physique*, because the reader of the *Science Allemande* is oriented towards looking for them and appreciating their importance. But without this 'study aid', the relevant material in the *Théorie Physique*, the opposition of intuitive *finesse* to deductive *géométrie* is liable to seem an anomaly of obscure origin for which very little in recent Anglo-Saxon philosophy can provide Duhem's readers with any kind of preparation. The background of such a reader is likely to be either in a tradition deriving from the later Wittgenstein for which all is *finesse* and rigorous deduction can have little place, or else he is likely to be trained in a school deriving either from logical positivism, or critics of that movement like Popper, for whom logic, or *géométrie* is all and *finesse* an embarrassment to be explained away. The preparation Duhem's early readers brought to the work was the turn of the century tradition of 'critique des sciences' to which his own earlier writings had made a significant contribution. The overlap between the earlier writings and the book was hardly calculated to alert readers to shifts in his viewpoint such as the new importance Duhem attached to common sense and *finesse*.

For *finesse* and Duhem's distinctive doctrine of natural classification play only a minor rôle in the papers of 1892–94 in the *Revue des Questions Scientifiques*. They only appear there in response to the criticism of Vicaire, and the evidence is that Duhem's *Commentaires* of the same years did not impress Duhem's contemporaries as the work of a man for whom common sense was the foundation of science. Picard refers to the issue several times. One quotation should make the point:[2]

> Duhem's energetics offered us above an example of the way he posited his principles *a priori*. This seeming pretension to divine nature troubled more than one reader of the *Commentaires*, worried by the minor rôle experiment seems to have played in the elaboration of the theory, and this physico-mathematical edifice was capable of provoking some by its at least apparent arbitrariness.

The *Commentaires* played a large part in the analysis of Abel Rey, probably the most thorough and careful Duhem's work received. Rey quoted from *Mécanique* Duhem's criticism, in the name of common sense, of Ostwald's metaphysics of matter-free energy,[3] before making this footnote comment:[4]

> This criticism is very strange coming from Duhem. Indeed as will be seen in the following chapter he tries to construct a *purely mathematical* physical theory, i.e. one without matter and that he rests this theory on principles relating to energy.

It is possible though that the "brutality"[5] of the *Commentaires* was more apparent than real, that the increasing rigour of late-nineteenth-century mathematics came as a shock to those not used to finding it in theoretical physics, but Duhem's training in mathematics was well suited to make him demand and provide such rigour in his own work.

I have already referred to the mistakes and errors of interpretation historians have been able to find in Duhem's work. I have suggested that these should be partially excused by the conditions and speed of working, but still, no matter how important the subject was to him, perhaps he should have recognized that he was taking on something beyond his powers and settled for more modest targets. It has often seemed to me that some alleged mistakes were not mistakes at all and that Duhem was subtler than his critics, but it has to be admitted that there is an element of '*hubris* normalien', about the attempt to change the course both of theoretical physics and the historiography of science all at once with no significant assistance from others. In some ways his historical researches remind me of that other product of the *École Normale*, the Pascal scholar Ernest Havet, who attempted, without any knowledge of Hebrew and at the end of a career largely devoted to teaching Greek literature, to rewrite the history of Judaism and Christianity. Only the superb confidence of the *Normalien* gained from membership of that élite intellectual community could have given either Havet or Duhem the overconfidence to undertake their respective self-imposed tasks. Duhem's preparation in Latin and Greek, despite his brilliance, did not go beyond what he got at school. Mistakes were inevitable. What is astonishing is the extent of his accomplishment, and despite the other obstacles discussed below, his work still plays a large rôle in the historiography of mediaeval science.

Duhem's work, then, offered some substantial hostages to fortune: shifts of viewpoint that did not always become explicit, and simple mistakes his speed of working gave him no time to correct. To these internal factors, external factors, relating to the preoccupations of the time, have to be added. I have already discussed how ideologically-motivated these were: little could be said or written that did not have direct political implications. This means that Duhem's readers divide up naturally according to their religious or anti-religious affiliations, irrespective of whether their main interests were in philosophy of science or in mediaeval intellectual history. An interest in the philosophy of science went with an interest in the history of science, but following Duhem into the Middle Ages required special abilities and competence unlikely to be found in most such readers. Indeed the most likely possessors of such abilities would be Catholic scholars involved in neo-Scholasticism. Below I consider first the philosophy of science before proceeding to its history.

2. The Partial Reception of the Philosophy

Duhem's philosophical writings aroused interest right from the beginning, and were influential. Part of the evidence for this is a series of footnotes in the *Théorie Physique* mainly concerned to protect Duhem's priorities and distinguish his position from others in a continuing debate. In 1936 Hélène claimed that deliberate non-citation of Duhem's work was common:[6]

> "This young man will never teach in Paris," declared Berthelot. These words had the effect of a sentence. Then began that thirty-year struggle between the Sorbonne on the one side and Pierre Duhem on the other. He will be the enemy, the man never to be spoken of, all of whose productions will be ignored, whose discoveries will all go unmentioned, whom by this silence and oblivion they will hope to discourage, whom even today they affect not to cite, even when a sentence in a work seems to be taken verbatim from one of his books.

There seems to be a problem of this sort here. In *Théorie Physique* ii/iv/i pp. 217–18 Duhem refers to his 1894 paper 'Physique Expérimentale' and notes its citation by Gaston Milhaud in 1895–96 and 1898, by Édouard le Roy in 1899 and 1900, and by the Sillonniste and Stanislas former pupil Joseph Wilbois[7] in 1899. In Section ii, attacking one of Poincaré's positions in the latter's *Valeur de la Science*, he remarks that "Mr. H. Poincaré knows that there is a possible objection to the theory he upholds; . . . "and notes:[8]

> In any case there is no reason for astonishment if it is noted that the above doctrine was published by us, in almost identical terms, as early as 1894, while Mr. Poincaré's article appeared in 1902. Comparing the two articles will convince the reader that in this passage Mr. H. Poincaré is attacking our way of thinking quite as much as Mr. le Roy's.

Le Roy was perhaps an easier target because of the greater generality of his position, but it seems in any case that Poincaré found it best to attack under that name Duhem who in turn, as remarked above, was compelled to assert his priority. In the following chapter,[9] he again refers to the citation of his earlier paper by Milhaud in 1896 and 1898.

The 1894 paper seems to have caused something of a sensation and found a ready audience for its ideas, and it is not difficult to see why. As remarked above, it is likely that the main background to this audience was the contemporary 'critique des sciences' tradition that grew up in the last quarter of the nineteenth century, and in some ways paralleled the German-language tradition of which Ernest Mach is perhaps the best-known representative. It seemed to cross the religious/anti-religious battle lines and might have continued to do so but for the Dreyfus affair. It seems to have originated in the work of Émile Boutroux and Henri Bergson,[10] and by the early 1890s it became possible for the influential editor of the *Revue des Deux Mondes* Ferdinand Brunetière (who ended his days a Catholic) to talk of the "bankruptcy of science" in his famous 1895 article 'Après une visite au Vatican'.[11] In contrast to the triumphalist scientism that had previously prevailed in intellectual circles, the new theme was of the limitations of science. It was once more O.K. to do metaphysics, and in 1893 the Jews Xavier Léon and Élie Halévy founded the *Revue de Métaphysique et de Morale*. The former remained as editor over many years as its stature grew and Catholics like Maurice Blondel, Édouard le Roy, and Victor Delbos became prominent contributors alongside non-Catholics like Henri Poincaré, Léon Brunschvicg, Abel Rey, and Bertrand Russell.

In that environment Duhem's 1894 article could serve, apart from his other ideas and commitments, as a penetrating analysis, unequalled elsewhere, of the limitations of science, an analysis from which writers like le Roy could sometimes draw, as Duhem remarked, "conclusions that go beyond the limits of physics."[12] But that of course would add to the journal's fame. In the *Théorie Physique* Duhem merely added to this article, rather than rewriting it, and so made it easy for readers to miss the shifts I have discussed above. Thus in 1904 Abel Rey had an apology to make:[13]

> This study was completed before the appearance of Mr. Duhem's articles in the *Revue Philosophique* (*sic*), articles synthesizing Mr. Duhem's views on physics. This explains why there are no references here to these articles, *which in any case change nothing in their author's general position*. (emphasis mine)

Writing here when publication of the *Théorie Physique* in that form had hardly begun, it is unlikely that Rey revised that judgement when it was complete and unlikely that he ever examined Duhem's final version carefully, for when in 1907 Rey's thesis was published as *La Théorie Physique chez les Physiciens Contemporains*,[14] what it has to say on Duhem is largely based on the 1904 article which it often reprints verbatim. Where he revised his judgements on Duhem's metaphysical commitments, the revisions are based on Duhem's 'Physique de Croyant' of 1905, not on the book.

Rey's treatment of Duhem was not only extremely thorough and careful, but also highly influential: it may even have determined the main emphases in all subsequent commentary on and use of Duhem in at least the Anglo-Saxon philosophical world, if not in Marxist circles as well. It would need much research to fill out and establish this conjecture, which depends on the well-known influence of the Vienna Circle on postwar Anglo-Saxon philosophy and on the use Lenin made of Rey's work in his *Materialism and Empiriocriticism* of 1909.[15] In Rey's eyes writings like those of Duhem had brought about a crisis of science, a crisis in which it seemed that it was no longer possible to extract from science a metaphysical account of the world of the kind positivists had hoped for. It seems that this concern found a response in both quarters, both in Lenin's concern to rehabilitate materialism in the face of the rise of idealism (and he took Duhem to share Mach's idealism), and in the concern of the early members of the Vienna group to rehabilitate positivism in the face of the new rigorous criticism of science. Preserving the usefulness of science as a bulwark against religion was no doubt another common concern.

The Vienna story is perhaps the more interesting and relevant. The *Vienna Circle Manifesto* of 1929[16] repeats at several points the debt of its members to their intensive reading of Duhem (no doubt under the influence of Mach) and this account is in agreement with Philipp Frank's in his 'Introduction: Historical Background' to his *Modern Science and its Philosophy*. As he tells us, Frank in effect founded the movement with Hans Hahn and Otto Neurath in 1907. He tells us that apart from Mach and Poincaré, they also spent a lot of time reading Rey's work, giving the clear impression that it was from the latter that they got their view of the problem situation, the rescue of positivism from the collapse of mechanistic science. It also looks as though they read Duhem through Rey's analysis.[17]

The distortions arising from this approach can well be imagined. Without common sense, Duhem's analysis becomes, as shown above, a radical scepticism whose only purpose could be to prepare the ground for the neo-Thomism of which Frank, apparently on hearsay, believes Duhem an adherent. Frank may well be the main source of the neo-Thomistic interpretation of Duhem this essay has spent so much time refuting.

3. Catholic Criticism of Duhem

In Chapter III I explored the reasons for Catholic resistance to the Duhemian approach to the philosophy of science. In brief the problem was its effect on natural theology. A sceptical view of the probative powers of science combined with the doctrine of the autonomy of physics made physics impossible to use apologetically and therefore also imperilled Catholic claims to authority. Though Lenin assumed the contrary in his *Materialism and Empiriocriticism*, scepticism as to the claims of science was welcome neither to the Catholic hierarchy nor to much Catholic opinion. Thus, as remarked above, Eugène Vicaire[18] could attack Duhem's 1892 papers for containing the "poison of scepticism", as dangerous now as in the writings of David Hume, and go on to claim that it was bad enough in the writings of the Berlin physicist Gustav Kirchhoff, but infinitely worse in a Catholic journal. Harry W. Paul has given a useful survey of Catholic criticism of Duhem.[19] For my purposes however it seems best to concentrate on Jacques Maritain, writing in 1932 when Duhem was 16 years dead.

Maritain criticizes Duhem in three different places, two of them extended footnotes, and the other a long digression. There are also a number of neutral references to him. A curious feature is his use of indirect citation. There is one reference to a specific page of Volume i of the *Système*,[20] and one likewise to the *Théorie Physique*. One other quotation[21] "a physical theory is not an explanation, but a system of mathematical propositions" from the latter does not get a page number when one could have been given.[22] The remaining references are to Picard,[23] and the quotation just cited is preceded by the remark as "... Émile Picard recalled in his lecture of 16 December 1929 to the Académie des Sciences . . ." thus giving the impression that Maritain has not read the *Théorie Physique* or seriously studied it, for on page 272 he refers to the same lecture of Picard for the view that:

> If it is asked what Fresnel's theory of waves reduces to, the reply must be, and here we are touching on a central point of the philosophy of science, a system of *differential equations*. . . . More or less similar circumstances occur elsewhere and in these conditions it will be understood how difficult it is to condemn definitively the initial conception of a theory.

Thus using Picard as an authority for views examined at length by Duhem and known to Picard, in a manner reminiscent of Poincaré's method of non-citation. In such circumstances we cannot expect much critical subtlety.

On page 84, immediately following the passage cited in my Chapter V, Section 5, Maritain writes:

> In such sciences the rule of explanation (as Duhem saw very well) leaves aside the principles and physical causes . . . —but that does not stop the sciences remaining (as Einstein and Meyerson see very well) also physical, because it is in sensible nature that they have their conclusion.

At this point it is not particularly clear why Maritain thinks he is disagreeing with Duhem (or for that matter agreeing with Einstein and Meyerson), but disagreeing he is, for an extended footnote on that page goes out of its way to differentiate his position from Duhem's, and on page 125 a footnote beginning on page 123 tells us that:

> Let us add, to avoid any misunderstanding, that Σώζειν τὰ φαινόμενα [saving the phenomena] in no way implies the rejection of the search for causes and for explanatory hypotheses that Duhem on his side attributed to physical theory. . . . It is the very causal explanations and the figura-

> tive entities elaborated by the physical sciences that are arranged for *saving the phenomena* and they are *true* . . . to the extent that they succeed, without claiming to penetrate the very nature of things.

There is a reference back to pages 88–90 where Maritain tells us:

> It does indeed happen that in some of its chapters physics uses mathematical symbols, without then attempting *causal explanations* or constructing figurative hypotheses in virtue of which the mind might in some way dismantle the mechanism of the phenomena. But in truth when it so refrains, it is only that it cannot do otherwise, and that it has to make a virtue of necessity. Duhem's mistake was to seek in these rather exceptional cases, which he regarded as pure cases, the very type of physical theory. In reality they are limiting cases, where the mathematical transposition of the phenomena maintains itself for the moment all alone in the mind, without any underlying physical image, and they are so far from representing the very type of physical theory that the mathematical symbols they use are only waiting their chance to leave the domain of pure analytical forms and act as explicative entities.

On the factual plane this analysis has some plausibility, but it shows no sign of meeting Duhem's real concerns, among other things to offer a *logical* analysis of the consequences of following certain aims, not an account of how physics actually evolves, nor does it show any sign of engagement with Duhem's Pascalian ideal (shared with the mathematicians of his day) of rigorous logical deduction. But it can be seen that if the entities of physics really do behave in the way Maritain describes, they can be more easily subjected to the superior regulation of metaphysics, which is what Maritain wants, but at the price of an argument that has weakened the probative power of experiment even further than Duhem's position: what kind of experimental proof is compatible with using entities with the properties Maritain ascribes to them? By casting the problem in the form he did, postulating rigorous mathematical deduction of results, the whole to be checked (if appropriate) against metaphysics and experiment, Duhem made it possible to ask the basic questions of the probative power of both metaphysics and experiment in the most rigorous possible form. If favourable answers could not be extracted in these terms, then the possibilities on any other logically less clear terms become that much more remote, if that is, they can then be asked at all in any meaningful sense. Maritain has missed the point.

If that is what Maritain makes of Duhem's physics, in his favourite science of biology his failure to attend to Duhem leads him to misrepresent Duhem entirely. On Page 385 a footnote tells us:

> We see how narrow was Duhem's identification of the Σώζειν τὰ φαινόμενα [saving the phenomena] with the pure translation of physical data into a system of mathematical equations apart from any search for 'causal explanation'. In the sciences we are now speaking of the mathematical translation of the phenomena, important as it can be, plays an entirely instrumental not formal rôle, and the search for 'causal' empiriological explanations . . . is preponderant.

Duhem never made the alleged identification. As in his earlier writings, his *Théorie Physique* had nothing to say about biology, which, as we learn from the *Science Allemande*, he did regard as an experimental science whose aim presumably was to save the phenomena. The restriction of the scope of the former work was explicit and deliberate.

Maritain's lack of attention to Duhem is revealing. As an avowed Thomist he contributed to the *Revue de Philosophie*,[24] with which Duhem was associated, and the independently-published version of the *Théorie Physique* was published in a series, the *Bibliothéque de Philosophie Expérimentale*, associated with that journal. His mishandling of Duhem is, it seems to me, indicative of the embarrassment Duhem caused the neo-Scholastic movement, an embarrassment even more visible in the reception of his historical works.

4. Reception of the Historical Work

As remarked above, few non-Catholics had the competence to follow Duhem into mediaeval intellectual history, so that the likely readership for all the works of Duhem's latter years was largely of Catholics committed to neo-Scholasticism. Exceptions included Albert Dufourcq who offered his services to Blondel as a possible editor of the *Annales* when Laberthonnière was silenced,[25] and A. Darbon whose review of the historical work for the memorial volume was reasonably intelligent.[26] As a result Duhem made, and continues to make to this day little dent on anti-religious denigration of the Middle Ages. Favaro dismissed Duhem's work as ill-founded neo-Thomism,[27] and the multi-volume *Histoire de la Nation Française* edited by Duhem's correspondent, the republican politician and historian Gabriel Hanotaux, assumed that the history of science began with Galileo.[28] Another complicating factor was that the historical work began in effect ten years later than the philosophical, and its tortuous history could made it difficult to grasp its importance, beyond the fact that it allegedly dispelled the legend of mediaeval darkness. Hence an analysis of its fate has to concentrate on the more limited audience, largely neo-Scholastic, of those competent to follow it.

That audience, besides being Catholic, consisted largely of priests, bound by the 'Oath against Modernism' to uphold the doctrines of *Pascendi*,[29] and as remarked above in Chapter III *Pascendi* very explicitly enjoined a Scholastic framework oriented towards the teachings of St. Thomas Aquinas for the whole of metaphysics and the sacred sciences. A reader of Duhem's historical work with that background found particularly steep obstacles in his way, faced as he was with a committed opponent of the doctrines he was committed to defending. He was not dealing with neutral 'objective' history, a fact that every commentator since Duhem's time has been only too well aware of and has had to deal with in their own way, as Henri Bosmans had to in 1921 when Duhem was five years dead.[30]

> Duhem had his way of understanding history. He never understood the history of science as it had been understood by a Montucla, a Chasles, a Cantor. He never understood it as the mere objective recital of the facts. An external preoccupation dominates him always, though without ever leading him to falsify the truth. . . .

Duhem's Catholic readers could not be so sure that he had not falsified the truth, and indeed were committed in advance to believing otherwise.

The embarrassment that was likely to result, and did result, should be plain. A major physicist whose philosophy of science runs counter to the approved philosophy is bad enough, but things look bleaker when he discovers all the good science in the Middle Ages that was supposed to be a historical desert *but* attaches these good things to trends *not* among those commended by the Encyclical. When he goes on to deny that the Scholastic philosophy there exclusively commended has any existence, being an illegitimate amalgam of Christianity and irredeemable paganism, the embarrassment is complete. There is little left for such readers but to behave like some of Duhem's anticlerical opponents and pass over his views in silence, or skate over them with the minimum of misrepresentation. I know of one Catholic other than Dufourcq who actually mentioned what Duhem said about Aquinas, the Jesuit Adhémar d'Alès, editor of the five-volume fourth edition of the *Dictionnaire Apologétique de la Foi Catholique*, who has this to say:[31]

> Saint Thomas does not conceal his borrowings and his text seems studded with references. This eminent expert on science could stop at this

> external observation and fail to recognize the existence of a Thomist philosophy

He cites Duhem[32] at this point and continues:

> This judgement betrays some precipitation. The truth seems otherwise if, instead of stopping at the source count, we take the trouble of feeling the thread of his development. We recognize at once that Saint Thomas transforms everything he touches.

Duhem's correspondent does not argue his case further.

The corresponding article in the *Dictionnaire de Théologie Catholique* avoids the issue by not mentioning Duhem's views at all. The article on Aquinas as a commentator on Aristotle,[33] by Duhem's correspondent Garrigou-Lagrange, confines itself to citing Grabmann implausibly extracting praise from Duhem's comments on Aquinas in *To Save the Phenomena*[34] in a passage whose tone contrasts nicely with Duhem's attitude:

> Everywhere Thomas walks the narrow path of truth, and to the utmost of his ability seeks light and clarity on the problems before him. He starts out from previous results, utilizes the conclusions already attained, adds proof to proof, observation to observation, until the solution sought stands out clearly. Everywhere he separates real from apparent knowledge, the certain from the probable, definite conclusions from hypotheses. Pierre Duhem, historian of the Copernican system, considers it the high merit of Thomas to have taken the following stand in regard to the Ptolemaic astronomy. . . . The hypotheses on which an astronomical system is based do not become demonstrated truths by the very fact that their deductions agree with observations.

Duhem could hardly be praising Aquinas for repeating what he had learned from Averroes.

The most interesting case, however, is Étienne Gilson, like Maritain a leading figure in inter-war neo-Scholasticism, and whose interests, following early work on Descartes, have centred primarily on mediaeval intellectual history. His *History of Christian Philosophy in the Middle Ages*, a well-known source, is by no means as long as Duhem's *Système*, but it covers much the same ground and will do for comparison. His text discusses Duhem in a number of places, but always in relation to the *Études*, and on aspects of the history of science. His notes cite Duhem more often and the *Système* is usually cited where relevant, once indeed with approval[35] when he explicitly agrees with Duhem's positive assessment of the orthodoxy of John Scot Eriugena,[36] thus showing his awareness of the theological dimension of Part iii (Tome iv and v) of the work. From Volume v he cites the chapter on Albertus Magnus,[37] but not the following chapters on Thomas Aquinas and Siger of Brabant. In his independent work on Aquinas he cites Duhem once only, a 1909 article in the *Revue de Philosophie* 'Du Temps où la Scolastique Latine a connu la Physique d'Aristote'.[38] In view of Gilson's reputation for careful scholarship, the omission can hardly be accidental or an oversight—it does not require much careful scholarship to discover that hundred-page chapters on Aquinas and Siger of Brabant follow that on Albertus Magnus. In contrast to Duhem's view of Aquinas's philosophy as an ill-fitting jigsaw puzzle cobbled together by a scholar with more loyalty to his faith and his sources than critical sense, Gilson remarks that[39] "The powerfully unified character of the doctrine will certainly have been noted or at least felt: it constitutes a total explanation of the universe from the point of view of the reason." It is altogether incredible that Gilson did not *know* of Duhem's opposite view: it seems he has decided that the best way out of his difficulty was to pass over the disagreement in silence. I may add that Copleston's *History of Philosophy* follows the even safer policy of only citing Duhem in relation to the late Middle Ages.

5. Conclusion

The upshot of all this is that Duhem seems to have fallen between every available stool and must be as frustrated in heaven as he was on earth. Excluded from academic advancement by politics as a result of his quarrel with Berthelot and the Dreyfus affair, he set to working all the harder, teaching and researching. His scientific work was rewarded, though not with many readers, with non-resident membership of the Académie, but his opinion of that honour is adequately expressed by the following quotation from a letter to his daughter:[40]

> You tell me that I've had more influence since I've been a Correspondent. I think the reverse is true: more and more my works have passed unnoticed. This year *one* copy of my great treatise on electricity has been bought. For me this honour has the value of a crown placed on the coffin in which the gentlemen of physics have nailed me still alive.

The later historiography of energetics does not seem to have changed this situation and we still lack a good critical account of the rôle of Duhem in these developments.

Surely, it might be said, the philosophy had a different fate? The *Théorie Physique* was translated into German under the auspices of Ernest Mach[41] and into English in the 1950s,[42] and has spawned an extended literature, but in a fate not unlike that of Pascal's *Pensées*, that literature treats it as the product it is not of scepticism and instrumentalism. In France there were no reprints between 1913 and the mid-1980s, because, or so it seems to have been felt until recently, positivism has been left behind. The German edition has, however, recently been reprinted.

Perhaps, then, Duhem's historical work reveals a different story? Hardly: it is famous for arguments more peripheral than central, and misunderstood anyway, while its central theses have been ignored. Astonishingly the *New Catholic Encyclopedia* gives Gilson, not Duhem, the credit for rejecting the view that there was one common scholastic synthesis in the Middle Ages—a classic case of the kind of non-citation discussed above.[43] Duhem remains the pioneer in all historical researches aimed at unearthing the real contribution Christian belief made to the rise of modern science and philosophy. Such researches have been a major academic industry over the years, but in none of them are Duhem's contributions given due weight or the benefit of proper critical assessment. I cannot help but feel that scholarship is thereby the poorer.

So we find Duhem, as a Catholic ignored and rejected by republicans, with the exception of École Normale contemporaries and foreigners, and likewise ignored by Catholics because his views were suspect—apart from those in no position to advertise their agreement with Duhem except at risk to themselves. Duhem's isolation was thus well-nigh total, almost as total as that of Laberthonnière. It was not only the physicists who had nailed him alive in his coffin.

But Duhem had the faith to take even that. Long ago he had used against Berthelot a quotation from the *Imitation of Christ* of Thomas à Kempis—he kept a copy by his bedside:[44]

> Tell me—where now are all those teachers and masters whom you knew well while they yet lived, and were eminent in learning? Already others hold their positions, and I know not whether they think back on them. In their lives they seemed to be something, but now there is no word of them.

The quotation has more than an echo of Villon's "Où sont les neiges d'antan?" (Where are the snows of yesteryear?) and also of course applied to him as well as to the contemporaries he had so damagingly quarrelled with. But as he had said the previous year at the conclusion of an article in the *Revue des Questions Scientifiques* ('L'Évolution des Théories Physiques du XVIIe Siècle jusqu' à nos jours'):[45]

> In the immense labour, there is no worker whose work has been lost. Not that that work has always served the purpose its author intended: the rôle it plays in science often differs from the rôle he attributed to it; it took the place marked in advance by Him who controls all this agitation.

That quotation is as suitable for closing this study as it was for closing his daughter's biography.

Notes

Works are cited by author and short title. Full details are given in the bibliography which follows. Except in the cases listed below, works without a stated author are by Duhem. In addition the following abbreviations are frequently used:

Annales	*Annales de Philosophie Chrétienne.*
Bordeaux	Société des Sciences Physiques et Naturelles do Bordeaux, *Mémoires* Vol i part i (1917) and vol i part ii (1927).
DSB	*Dictionary of Scientific Biography,* ed. Gillispie and others, New York: Scribners, 1970–1980.
Laf	Pascal, *Pensées* cited by the number in editions deriving from that of Louis Lafuma (including the translations of Krailsheimer [Penguin 1966] and Warrington [Everyman 1973]).
NCE	*New Catholic Encyclopedia,* New York: McGraw-Hill, 1967.
Phenomena	ΣΩΖΕΙΝ ΤΑ ΦΑΙΝΟΜΕΝΑ: *Essai sur la notion de Théorie physique de Platon à Galilée*
RQS	*Revue des Questions Scientifiques.*
Théorie	*La Théorie Physique, son Object et sa Structure.*

Citations of the latter are in the form *Théorie* part/chapter/ section/ (page no. of 2nd edition 1914). Other abbreviations should be self-explanatory. Lower-case Roman numerals are used for volume and chapter numbers as appropriate. Unless otherwise stated, letters to Duhem are now in the Duhem archives now at the Academie des Sciences, Paris. I thank Dr. Donald G. Miller for copies. Duhem's letters to Blondel are at the Archives Blondel, Université Catholique de Louvain, Collège Thomas More, Chemin d'Aristote 1. B1348 Louvain-la-Neuve, Belgium. I thank Mme. G. Mosseray for copies and for commenting on my transcriptions, and Mr. Duncan McGibbon for discussing them with me. The Tannery correspondence is cited from Tannery, *Mémoires Scientifiques* vol. xiv. Unless otherwise indicated all translations are mine.

CHAPTER I

1. For an updated version of Duhem's own list see *Bordeaux* 41–70 and 607–634. Jaki, *Uneasy Genius*, 437–455 has a fuller list and D.G. Miller of the University of California Lawrence Livermore Laboratory a yet fuller though unpublished one.
2. Duhem, 'Notice' in *Bordeaux,* 72
3. For the relevant details see *Le Centenaire de l'École Normale.*
4. In 1961. See Duhem, *Hydrodynamique.*
5. In O'Rahilly, *Electromagnetic Theory.*
6. Hadamard, 'Œuvre de Duhem dans son aspect mathématique'.
7. See e.g. Harding, *Can Theories be Refuted*?
8. Entitled *Mediaeval Cosmology.*
9. By Philipp Frank (1912) and Michael Cole (1980) respectively.

10. By E. Dolan and Chaninah Maschler (1969). A Portuguese translation has also been published (*Salvar os Fenomenas*).
11. In 1981, Friedrich Adler's German translation of 1948 was reprinted in 1978. There is also now an Italian translation, *La Teoria Fisica*.
12. The works of Roberto Maiocchi and Anastasios Brenner are honourable exceptions.
13. See for example Favaro, 'Galileo'; Frank, *Modern Science*, 'Introduction'; Nye, 'Moral Freedom'; and Paul, *The Edge of Contingency*.
14. 'Thermochimie'.
15. Duhem to Blondel, 12 January 1896.
16. Bordeaux, 71.
17. See Paul, 'Crucifix' and *Edge of Contingency*.
18. See especially his *DSB* article.
19. Jaki, *Uneasy Genius*.
20. Jordan in Bordeaux, 3–40; H. Pierre-Duhem, *Savant*.
21. Collingwood, *Autobiography*.
22. 'Notice' in *Bordeaux*, 71–169.
23. Lowinger, *Methodology*.
24. Maiocchi, *Chimica*.
25. Brenner, *Science*.
26. On this incident see Jaki, *Uneasy Genius*, 181–82 and Paul, 'Scholarship and Ideology'.
27. On this see Dupuy, *École Normale*, and Anon, *Centenaire*.
28. See e.g. H. Pierre-Duhem, *Savant*, 126–152; Paul 'Crucifix and Crucible'
29. See the testimony of Jordan, and Hélène herself has much to say about Duhem's peaceable relationships with Jewish and Protestant friends. See e.g. H. Pierre-Duhem, *Savant*, 141.
30. A figure of this order seems implicit in the class lists in *Centenaire*.
31. Lakatos, 'History of Science'.

CHAPTER II

1. See his *Conjectures,* Cap. ii, 66–96.
2. Bury, *France* is a convenient account of nineteenth-century French history.
3. H. Pierre-Duhem, *Savant,* 126.
4. On the effects of such policies on the Catholic community see the writings of H. W. Paul, particularly his 'Crucifix and Crucible', 'Quest of Kerygma' and 'Bankruptcy of Science'.
5. See in addition to Bury, *France,* Cap. xi, 170–79, the fascinating account of A. Dansette, *Boulangisme.*
6. See H. W. Paul, *Second Ralliement*; E. R. Tannenbaum, *The Action Française*; Eugen Weber, *Action Française.*
7. H. Pierre-Duhem, *Savant*, 130–31 and Paul, *Crucifix*, 202–211.
8. H. Pierre-Duhem, *Savant*, 126.
9. The following account largely corresponds with E. Poulat *Catholicisme*, *Démocratie, et Socialisme.*
10. On this see Vidler, *Variety*, 191–220.
11. H. Pierre-Duhem, *Savant*, 126.
12. Jaki, *Uneasy Genius*, 90.
13. Testis, 'Semaine Sociale' plus others under the same pseudonym, an article by Laberthonnière, two letters, and a reply from the Spanish Jesuit Pedro Descoqs.
14. 'Vertus allemands'. See also my 'Trouble with Authority'.
15. On Berthelot see also Paul, 'Crucifix and Crucible' and 'Bankruptcy of Science', and Crosland, *DSB qv.*
16. H. Pierre-Duhem, *Savant*, 158–169.
17. On positivism see Gouhier, *Jeunesse d'Auguste Comte*; Hayek, *Counter-Revolution of Science*; Simon, *European Positivism*, in addition to Maiocchi, *Chimica.*
18. Gibbon, *Decline and Fall*, Chapter ii.
19. Maiocchi, *Chimica.*
20. 'Notation Atomique'.

21. Vicaire, Valeur Objective'.
22. Popper, *Conjectures and Refutations*, 231–34.
23. 'Physique Expérimentale'.

CHAPTER III

1. 'Physique de Croyant' in *Théorie Physique*, 428.
2. 'Physique de Croyant' in *Théorie Physique*, 435.
3. *Annales* cli (1905-06), 44–67 and 133–159; reprinted in *Théorie Physique*, 2nd edition, 413–472.
4. Rey, 'Philosophie Scientifique'.
5. Rey, *Contemporains*, seemingly completed in 1905.
6. On this incident see Paul, 'Bankruptcy of Science'.
7. Blondel, 'Apologétique', reprinted in Blondel, *Premiers Écrits, 5–95.*
8. Duhem to Blondel, 17 January 1897.
9. Duhem to Blondel, 5 January and 27 December 1893; Blondel to Duhem, 17 August 1893.
10. For the text of Duhem's interventions see *Congrès Scientifique International des Catholiques, Compte Rendu* i, 313–15 and 322–25.
11. *Actes de Pius X* (1908), 84–181; English translation *Encyclical Letter . . .* 1907.
12. See McCool, *Catholic Theology* for a full discussion.
13. Mentré, 'Duhem'and Paul, *Contingency,* 157–160.
14. Vicaire, 'Valeur Objective', 453; Paul, *Contingency,* 165.
15. This account is largely based on McCool, *Catholic Theology*, andThibault, *Savoir et Pouvoir.* For a hostile contemporary accountseealso Chapter x (by L. Laberthonnière) of Lecanuet, *Vie de l'Église sous Léon XIII*, 447f.
16. Leo XIII, *Allocutiones* i, 88–108, translation (apparently by Gilson) in Maritain, *St. Thomas Aquinas*, 183–214.
17. On this see Gouhier, *Cartésianisme et Augustinisme,* caps iii, iv, and v.
18. See Knowles, *Evolution of Mediaeval Thought.*

19. See Bulliot, 'Faut-il changer l'Orientation de la Néo-Scolastique?'
20. Duhem to Tannery, 3 July 1897.
21. See Paul, *Edge of Contingency* 164f.
22. Maritain, *Distinguer pour unir.*
23. Aristotle, *Analytica Posteriora.*
24. See Westfall, *Construction.*
25. See Poulat, *Crise Moderniste;* Vidler, *Modernist Movement,* and *Variety.*
26. Paul, *Edge of Contingency,* 177–78; H. Pierre-Duhem, *Savant*, 104–05; *Revue de Philosophie* xix (1919), 458.
27. *Revue de Philosophie* iii/i, 250.
28. See Blondel-Valensin i, 223; Blondel-Wehrlé i, 300–01; Duhem to Blondel, 22 July 1905; Laberthonnière to Duhem, 16 August and 9 October 1905.
29. When it was put on the *Index* and its editor Laberthonnière forbidden to publish. See *Index Librorum Prohibitorum* (1948 edition).
30. Denis (review of F. Picavet, *Esquisse*) *Annales* cl (1905), 73–83.
31. *Annales* cl (1905), 301 and 302–05.
32. *Annales* cli (1905–6), 5–31.
33. Blondel-Wehrlé i, 324–26.
34. 'Notre Programme', 29.
35. 'Notre Programme', 23–25.
36. *Annales* civ (1907–08), 5–9.
37. *Annales* cli (1905–06), 535–37.
38. *Index Librorum Prohibitorum* (1948 edition).
39. Hooykaas, *Religion and the Rise of Modern Science.*
40. Duhem to Blondel, 12 January 1896.
41. H. Pierre-Duhem, *Savant*, 61.
42. See in addition to the published reports *op. cit.* (note 10); H. Pierre-Duhem, *Savant,* 157–58; A. Gardeil, *Revue Thomiste* (1894), 569–585, 738–759; A. Baudrillart, *Vie*

i, 552; P. Pisani, *Correspondant*, 25 September 1894; Lecanuet, *Vie*, 313–14.

43. Duhem to Blondel, 25 July 1905.
44. Laberthonnière to Duhem, 31 March, 14 August, and 13 November 1913. Laberthonnière had also commissioned a general survey piece on Duhem's work from Viscomte R. d'Adhémar.
45. Blondel to Duhem, 16 July 1913; Vidler, *Variety*, 84.
46. Hume, *Treatise* Bk. ii, Cap. ii, Sec. ii, 417.
47. *Bordeaux*, 15.
48. Duhem to Blondel, 20 July 1913. Extracts from this letter are printed in Blondel-Wehrlé (i, 531) and Lecanuet, *Vie* 478–79). The relevant chapters (ix, x, and xi) of the latter were in fact written anonymously by Laberthonnière: see Louis Canet's note in L. Laberthonnière, *Philosophie Personnaliste*, 426. I thank the late Rev. Dr. Charles Duthie for a sight of his copy of this work.
49. *Phenomena*, 136.
50. H. Pierre-Duhem, *Savant*, 203. Duhem's views on Aquinas will be explored further in Chapter IX.

CHAPTER IV

1. Picard, *Vie*, 50.
2. Strowski, 'Secret' 791.
3. Thirion to Duhem, 26 October 1907.
4. See 'Archimède' and 'Mersenne'.
5. Maire, *Œuvre Scientifique*
6. Maire, *Bibliographie Générale*.
7. On editions of Pascal see Louis Lafuma, *Histoire*, and C. S. Duthie, *Pascal's Apology*.
8. See Krailsheimer, *Pascal* and 'Introduction' 11–29 to his translation of the *Pensées*.
9. *Œuvres Complètes*, 221–25.

10. Duhem's view. See 'Mersenne'.
11. *Œuvres Complètes*, 199–221.
12. *Ibid.*, 230–32.
13. *Ibid.*, 348–355 and 355–59.
14. *Ibid.*, 371–468.
15. *Œuvres Complètes*, 657.
16. By Filleau de la Chaise. See Dubois de la Cour (pseud.), *Discours*.
17. *Œuvres Complètes*, 349–350.
18. Popkin, *History of Scepticism*.
19. Schmitt, *Cicero Scepticus*.
20. Montaigne, *Essais* ii/xii, in Montaigne, *Œuvres Complètes*, 415–589; Popkin, *History* iii, 42–65.
21. Aristotle, *Analytica posteriora* ii/xix, 99b15–100b7.
22. Leibniz, *Philosophische Schritten* i, 381–82; partial translation in G.W. Leibniz, *Discourse on Metaphysics*, 129–131.
23, Foucher, *Critique*.
24. Arnauld, *Des Vrayes et des Fausses Idées*.
25. *Œuvres Complètes*, 356-57.
26. Arnauld and Nicole, *Logique* ive partie, 2e défaut.
27. Laf, 110.
28. *Ibid.*
29. Eastwood, *Revival of Pascal*.

CHAPTER V

1. Laf, 511–13; *Théorie* part i, Chapter iv.
2. *Bordeaux* i, 27–28.
3. *Théorie* i/iv/viii, 131–32.
4. H. Pierre-Duhem, *Savant*, 220.
5. In Petit and Leudet (eds.) *Les Allemands*.
6. 'Quelques Réflexions'.
7. Maiocchi, *Chimica*, 232–35.
8. *Théorie* i/iv/x, 153–54.
9. *Théorie* ii/vii/v, 397–400.

10. *Théorie* ii/vi/x, 329–32.
11. *Science Allemande,* 4–5.
12. *Ibid.,* 29.
13. *Ibid.,* 27f.
14. *Ibid.,* 28–29.
15. *Ibid.,* 29.
16. *Ibid.,* 76.
17. *Ibid.,* 56–60.
18. *Ibid.,* 77–78.
19. *Ibid.,* 78.
20. *Ibid.,* 78–81.
21. *Ibid.,* 82–84.
22. *Ibid.,* 84.
23. Laf, 406.
24. *Science Allemande,* 25.
25. *Ibid.,* 114.
26. *Ibid.,* 126f.
27. Picard, *Vie,* 32.
28. Owen, 'Phenomena'.
29. *Mixte,* 179–185.
30. Laf, 298.
31. Laf, 308.
32. Laf, 511.
33. Krailsheimer, 'Introduction' (23) to his translation of Pascal, *Pensées.*
34. *Œuvres Complètes*, 355.
35. *Ibid.,* 230–32.
36. Eastwood, *Revival of Pascal,* 52.
37. Lanson, 'Pascal', 21.
38. *Théorie* i/i/ii-iii, 6–17.
39. *Théorie* ii/ii/i-ii, 179–190, esp. 184.
40. See especially 'Croyant', sec. ix in *Théorie,* 462–472.
41. Lowinger *Methodology*, cap i.
42. Maiocchi, *Chimica*, 277f.
43. Simmons, 'Sciences, Classification of'.

44. *Phenomena*, 1–2.
45. Maritain, *Distinguer pour Unir*, 82.
46. *Ibid.*,
47. *Théorie* ii/iv/i, 217–18 (note).
48. *Œuvres Complètes*, 231.

CHAPTER VI

1. 'Physique de Croyant', 'Valeur de la Science', *Théorie*, 413–472 and 473–509 respectively.
2. *Théorie* i/ii/i, 23–26.
3. *Théorie* ii/iii/i, 197.
4. 'Réflexions'; 'Notation atomique'; 'Physique et Métaphysique'; 'École Anglaise'; 'Monde Inorganique'; 'Physique Expérimentale'.
5. *Théorie* ii/vi/ii, 278–285.
6. Milhaud, *Certitude Logique*, 109–110 fn, for discussion of Cornu's report to the Académie des Sciences.
7. 'Étude sur l'Œuvre de George Green'.
8. *Ibid.*, 256.
9. 'Réflexions', 146.
10. *Ibid.*, 147
11. *Œuvres Complètes*, 349.
12. 'Valeur Objective', 476.
13. Laf, 298.
14. Laf, 52.
15. *Théorie* ii/vi/x, 330.
16. Laf, 298.
17. *Théorie* i/i/v, 17–21.
18. Maiocchi, *Chimica*, 52–60.
19. *Théorie* i/ii/iv-v, 32–40.
20. *Théorie* i/iv, 77–154.
21. Maiocchi, *Chimica*, 145–156.
22. *Théorie* i/i/i, 3–4.
23. *Théorie* i/i/ii, 7.
24. *Théorie* i/ii/iv, 35.

25. Laf, 423.
26. *Théorie* i/iv/v, 153–54.
27. Laf, 93.
28. *Théorie* ii/vii/v, 397–400.
29. *Théorie* ii/i/iii–v, 163–178.
30. *Théorie* ii/ii/ii, 189.
31. See e.g. Popper, *Logic*, 81–82.
32. *Théorie* ii/ii/iii, 195.
33. *Théorie* ii/iii/iii-iv, 206–215.
34. *Ibid.*, 213.
35. *Théorie* ii/iv/i, 218–19.
36. Maiocchi, *Chimica*, esp. 178–190 and 197–206.
37. Popper, *Logic*, 110.
38. Harding, *Can Theories be Refuted?*
39. *Ibid.*, 118.
40. Popper, *Conjectures*, 243.
41. 'Étude sur l'Œuvre de George Green' especially concluding paragraph; also *Théorie* ii/vi/ii, 282.
42. *Théorie* i/i/v, 17–21.
43. *Théorie* i/ii/i, 26.
44. *Théorie* ii/vi/ii, 278–285.
45. *Théorie* ii/vi/iii, 285–89.
46. *Théorie* ii/vi/iv-v, 289–303.
47. *Théorie* ii/iv/i, 217–18.
48. *Théorie*, xvi.
49. *Théorie* ii/vii/i, 333–36.
50. Vicaire 'Valeur Objective', 479–484.
51. *Théorie* ii/vii/iii, 384–391.
52. *Théorie* ii/vii/iv, 391–94.
53. *Théorie* ii/vii/v, 394–407.
54. *Théorie* ii/vii/iv, 392.
55. *Théorie* ii/vii/v, 404–05.
56. *Ibid.*, 405–06.
57. *Ibid.*, 407.
58. *Théorie* ii/vii/vi, 408.

CHAPTER VII

1. *Théorie* ii/vii/ii, 336-384.
2. *Théorie* ii/vii/iii, 384-391.
3. Kuhn, *Essential Tension,* 66–104.
4. Maiocchi, *Chimica,* 277–291.
5. Ariew and Barker (1986).
6. Popper, *Open Society* and *Poverty of Historicism..*
7. *Œuvres Complètes,* 231.
8. On Pasteur see Geison, *DSB qv.*
9. *RQS* lvi (1904), 394; *Origines* i, 156.
10. *Théorie* ii/vii/vi, 408-09.
11. 'Aimantation par Influence', 'Étude Historique'.
12. Hannequin, *Essai Critique.*
13. Dühring, *Kritische Geschichte,* 1st ed., 10–11; 2nd ed., 9–10.
14. Lakatos, 'History of Science'.
15. Lakatos, *Proofs and Refutations.*
16, Kuhn, 'Notes on Lakatos', 143.
17. *Théorie* ii/vii/vi, 408.
18. *Bordeaux,* 158.
19. *Ibid.,* 169.
20. *Théorie* ii/vii/vi, 409.
21. 'Analyse', 269.
22. 'Analyse', 270.
23. Lakatos, 'Falsification' and 'History of Science'.
24. Mach's inscribed presentation copy is in the Science Museum, London.

CHAPTER VIII

1. *Mécanique,* Cap vii, 62–70.
2. *RQS* liv (1903), 462–516.
3. *RQS* liv (1903), 469; *Origines* i, 13.
4. See Rose and Drake, 'Pseudo-Aristotelian Questions'.

I thank the late Dr. C. B. Schmitt for this reference.
5. *RQS* iv (1904), 561; *Origines* i, 62.
6. Bosmans, 'Pierre Duhem'.
7. *Ibid.,* 40–41.
8. Tannery to Duhem, 5 December 1903.
9. *Origines* i, i.
10. *Ibid.,* ii.
11. There are references to Tartaglia and Jordanus in Montucla, *History* ii, 1st edn., 417 and 621 (2nd edn., 506 and 691) and in Cantor, *Vorlesungen* ii, 1st edn., 55 (2nd edn., 60).
12. Cited by Duhem (*Origines* i, 35) from *Ztsch. Völkerpsch. Sprachwiss.* xiv (1883), 386.
13. *RQS* liv (1903), 507; *Origines* i, 51.
14. *RQS* liv (1903), 465; *Origines* i, 8.
15. *Origines* i, ii.
16. Tannery to Duhem, 7 November 1903; Duhem to Tannery, 11 November 1903.
17. See e.g. Claggett, *Science of Mechanics.*
18. Duhem to Tannery, 5 July 1904.
19. *RQS* lv (1904), 560-61; *Origines* i, 61–62.
20. *RQS* (1903), 468-69; *Origines* i, 11–12.
21. *Ibid.,*
22. Lemonnier, 'Études de Pierre Duhem sur Léonard de Vinci'. I thank Dr. K. H. Vettman for this reference.
23. Berthelot, 'Manuscrits de Léonard'.
24. Lemonnier, *op. cit.*, 31.
25. See e.g. Grant *Physical Science.*

CHAPTER IX

1. *Phenomena,* 2.
2. See Heath, *Aristarchus of Samos.*
3. Published in book form as *L'Évolution de l'Astronomie.*
4. Mansion in *Congrès Scientifique . . .*
5. Passage discussed in just about any treatment of Copernicus

and the astronomical revolution. Translation in Rosen, *Three Copernican Treatises.*

6. 'Réflexions', 146.
7. Vicaire, 'Valeur Objective', 499.
8. Schiaparelli, *I Precursori.*
9. *Simplicit in Aristotelis Physicorum* ... ii/ii. See *Phenomena,* 9–11 Heath's translation (from *Aristarchus of Samos*) is reprinted in Cohen and Drabkin, *Source Book*, 90–91.
10. Maiocchi, personal communication, reports a copy with the date 1893 but no title page, and it is not recorded in the Catalogue of the Bibliothèque Nationale, despite being issued by the Paris publisher Gauthier-Villars. Perhaps it did not get beyond proof stage. Cited by Duhem ('Métaphysique', 71) as 1893 but (*Théorie* i/iii/ii, 55) as1903.
11. 'Métaphysique', 71.
12. 'Métaphysique', 75.
13. *Théorie* i/iii/ii, 55.
14. Mansion, 'Caractère Géométrique'.
15. Grisar, *Galileistudien.*
16. From Kepler, *Opera* ed. Frisch i (1858), 112–153.
17. Martin, Hypothèses'.
18. Steinschneider, 'Ibn Haïtam'.
19. Renan, *Averroès et l'Averoïsme.*
20. Berti, *Copernico.*
21. *Phenomena*, 1.
22. *Phenomena,* 136.
23. *Phenomena,* 23.
24. *Phenomena,* 22–23.
25. *Phenomena,* 18, citing Ptolémée *Composition Mathématique* ed. Halma ii, 374–75.
26. *Phenomena,* 37.
27. *Phenomena,* 38–39.
28. *Phenomena,* 39–40 citing Maïmonide, *Guide des Égarées* ed. Munk ii, 194–95.
29. *Phenomena,* 78.

30. Clavelin, 'Galilée',.
31. *Phenomena,* 136–37.
32. *Phenomena,* 140.
33. Clavelin, 'Galilée', 313.
34. Koestler, *Sleepwalkers*, e.g. 477.
35. *Phenomena,* 127–28.
36. Collingwood, *Autobiography*, Cap ii.
37. Laf, 406.
38. Lakatos and Zahar, 'Copernicus'.
39. Sarton and Tannery, 'Appel'.
40. *Études* iii. See also 'Précurseur Français' amd 'François de Meyronnes'
41. *Système* vi, Cap i, 3–122.
42. *Savant*, 165.
43. Système vii, 3.
44. Système v, 568.
45. *Bordeaux*, 39.
46. *Bordeaux*, 22.
47. Cp. the hostility to Greek thought of Laberthonnière, *Réalisme Chrétien.*
48. *Système* vi, 66–69.
49. *Système* vii–ix.
50. Gilson trans. in Maritain, *St. Thomas Acquinas*, 199–200.
51. *Système* v, 569–570.
52. *Savant*, 203–04.
53. Dufourcq to Duhem, Pentecost 1912.
54. 'Accélération'.
55. Oregius, 1st edn., 194; 2nd edn., 294.
56. See Oakley, *Natural Law*, 433–457.
57. See Knowles, *Mediaeval Thought.*
58. See e.g. Grant, *Physical Science.*
59. Neugebauer, *Exact Sciences*, 206 note *ad* 77.
60. *Système* vi, 728–29.
61. M. Pierre-Duhem, *Savant*, 165–66.
62. Duhem to Blondel, 20 July 1913.

CHAPTER X

1. *Bordeaux*, 19–20.
2. Picard, *Vie*, 29.
3. *Mécanique*, 29.
4. Rey, *Contemporains*, 91.
5. Picard, *Vie*, 26.
6. H. Pierre-Duhem, *Savant*, 146–47.
7. There are a few suggestive references to him in Caron, *Sillon*. In the Blondel-Laberthonnière *Correspondance Philosophique* there is a letter (210) from Pautonnier of the Collège Stanislas commending him to Blondel along with Le Roy and one Vincent.
8. *Théorie* ii/iv/ii, 227.
9. *Théorie* ii/v/i, 250.
10. See e.g. Nye, 'Boutroux Circle'.
11. On this see Paul, 'Bankruptcy of Science'.
12. *Théorie* ii/iv/ii, 218.
13. Rey, 'Philosophie Scientifique', 699.
14. Rey's preface is dated 1905.
15. See especially Chapter v, Section viii, 364–378.
16. Translation in Neurath, *Empiricism and Sociology* i, 299-318. Mrs. Neurath's account of its origin can be compared with Frank's in *Modern Science*, 38.
17. One suggestive minor detail is the citation of Duhem's 'Réflexions' of 1892, presumably at second hand, from the *Revue des Questions Scientifiques* with both date and volume number wrong as T. xxxvi 1897 (instead of xxxi 1892) which is impossible for T xxxvi corresponds to the latter half of 1894.
18. 'Valeur Objective', 453. See also Duhem's note 'Monde Inorganique', 123.
19. See Paul, *Edge of Contingency*, Chapter v, 136–178, but see my 'Darwin and Duhem'.
20. Maritain, *Distinguer pour Unir*, 123.

21. *Ibid.*, 88.
22. To *Théorie* i/ii/i, 24.
23. Picard, *Coup d'Oeil.*
24. With e.g. his 'La Science Moderne et la Raison'.
25. Blondel-Wehrlé, *Correspondance*, 492–93.
26. *Bordeaux*, 667–718.
27. Favaro, 'Galileo'.
28. The relevant volume is xiv covering *Mathématiques, Mécanique, Astronomie, Physique, et Chimie;* by Henri Andoyer, Pierre Humbert, Charles Fabry, and Albert Colson.
29. *Actes de Pius X* v (1927), 141–186, especially 163–68.
30. Bosmans, 'Pierre Duhem' 447.
31. D'Alès, 'Thomisme', in d'Alès (ed.), *Dictionnaire Apologétique* iv (1922), col. 1693–94.
32. *Système* v, 569.
33. R. Garrigou-Lagrange, 'Thomas d'Aquin (*Saint*) le Commentateur d'Aristote', *Dictionnaire de Théologie Catholique* xv/1 (1946), col. 645.
34. *Phenomena*, 47; Grabmann, *Thomas Aquinas,* 35.
35. Gilson, *Christian Philosophy*, 613.
36. *Système* v, 61–65.
37. Gilson, *Christian Philosophy,* 668.
38. Gilson, *Thomisme,* nouvelle édition, 17.
39. Gilson, *Thomisme,* 5th edn., 497.
40. *Savant*, 170–71.
41. Duhem, *Ziel und Struktur.*
42. Duhem, *Aim and Structure.*
43. Weisheipl, 'Neoscholasticism and Neothomism', *NCE* x (1967), 337.
44. H. Pierre-Duhem, *Savant,* 155; quotation from Blaiklock's translation of Book i, Cap iii, Verse iv.
45. 'Évolution des Théories Physiques', 499, cited H. Pierre-Duhem, *Savant,* 237-38.

Bibliography of Works Cited

Agassi, J. *Towards an Historiography of Science, History and Theory,* ii, The Hague: Mouton, 1963.

Alès, Adhémar d', S.J. *Dictionnaire Apologétique de la Foi Catholique, contenant les preuves de la vérité de la religion et les réponses aux objections tirées des sciences humaines,* quatrième édition entièrement refondue, sous la direction de Adhémar d'Alès . . . Avec la collaboration d'un grand nombre de savants Catholiques. 5 vols, Paris, 1909–11, 1911–15, 1916–22, 1922–28, 1931.

———. 'Thomisme', *Dictionnaire Apologétique,* iv, 1922, cols 1667–1713.

Anon. *Le Centenaire de l'École Normale (1795–1895),* Paris: Hachette, 1895.

Anon. 'Société de Saint-Thomas d'Aquin', *Revue de Philosophie,* iii/i, 1903, 250.

Anon. *The New Catholic Encyclopedia,* 15 vols, New York: McGraw-Hill, 1967.

Ariew, R., Barker, P. *Philosophy of Science Association, Proceedings,* i, 1986, 145–156.

Aristotle. *Analytica Posteriora,* any edition.

Arnauld, Antoine. *Des Vrayes et des Fausses Idées, contre ce qu'enseigne l'auteur de la Récherche de la Vérité,* Cologne, 1678.

Baudrillart, Alfred. *Vie de Mgr. d'Hulst,* 2 vols, Paris: de Gigord, 1912.

Ben Maimoun, Moïse. *Le Guide des Égarés, traité de Théologie et de Philosophie, Public pour la première fois dans l'original arabe et accompagné d'une traduction française et de notes critiques, littéraires et explicatives par S. Munk,* 3 vols, Paris, 1856–57.

Berthelot, P. E. M. 'Les Manuscrits de Léonard de Vinci et les Machines de Guerre'. *Journal des Savants,* (February 1902), 116–120.

Berti, Domenico. *Copernico e le vicende del sistema copernicano in Italia nella seconda metà del secolo XVI e nella prima del secolo XVII,* Roma, 1876.

Blondel, Maurice. *L'Action, Essai d'une critique de la vie et d'une science de la pratique,* Paris, 1893; republished Paris: PUF, 1950, 1973.

———. *Les Premiers Écrits de Maurice Blondel, Bibliothèque de philosophie Contemporaine,* Paris: PUF, 1956.

———. 'Lettre sur les Exigences de la Pensée Contemporaine en matiére d'apologétique et sur la Méthode de la Philosophie dans l'etude du problème religieux', *Annales de Philosophie Chrétienne,* (January to July 1896) republished in Blondel, *Premiers Écrits, 5–95.*

Blondel, Maurice. Wehrlé, Joannès. *Correspondance, extraits,* annotations de Henri de Lubac, S.J. 2 vols, Paris: Aubier-Montaigne, 1969.

Bosmans, Fr. Henri. 'Pierre Duhem (1861–1916); Notice sur ses travaux relatifs à l'histoire des sciences', *Revue des Questions Scientifiques,* lxxx, (1921), 30–62, 427–448.

Brenner, Anastasios. *Duhem, Science, Réalité et Apparence, Mathesis,* Paris, Vrin, 1990.

Bulliot. Jean. 'Faut-il Changer l'Orientation de la Néo-Scolastique?', *Revue de Philosophie,* xiii (1913),46–56.

Bury, J. P. T. *France, 1840–1940,* 4th edn., London: Methuen, 1976.

Cantor, Moritz. *Vorlesungen über die Geschichte der Mathematik,* 4 vols, Leipzig. 1st edn., 1892, 2nd edn., 1900.

Caron, Jeanne. *Le Sillon et la Démocratie Chrétienne 1894–1910,* Collection *Histoire et Mentalités* dirigée par Robert Mandrou, Paris: Plon, 1967.

Claggett, *The Science of Mechanics in the Middle Ages,* Madison, Wisconsin: University of Wisconsin Press, 1959.

Clavelin, Maurice. 'Galilée et le refus de l'equivalence des hypothèses", *Revue d'Histoire des Sciences,* xvii (1964), 305–330.

Cohen, M. R., and Drabkin, I. E. *A Source Book of Greek Science,* Cambridge, Massachussets: Harvard University Press, 1958.

Collingwood, R. G. *Autobiography,* Oxford: Oxford University Press, 1938.

Copernicus, Nicolaus. *De Revolutionibus Oribium Coelestium,* Nürnberg, 1543.

Copleston, Frederic, S. J. *History of Philosophy,* London: Search Press, 1946.

Crosland, M.P., 'Berthelot, P. E. M.' *Dictionary of Scientific Biography* ii (1970), 63–70.

Daly, Gabriel. *Transcendance and Immanence,* Oxford: Oxford University Press, 1980.

Dansette, Adrien. *Du Boulangisme à la Révolution Dreyfusienne: Le Boulangisme 1886-1890,* Paris: Librairie Académique Perrin, 1938.

Denis, Charles. *Annales de Philosophie Chrétienne,* ci (1905), 73–83.

Descoqs, Pedro. 'Monophorisme et action Française', *Annales de Philosophie Chrétienne,* cix (1910), 225–251.

Dubois de la Cour (pseudonym = Filleau de la Chaise). *Discours sur les Pensées de M. Pascal,* Paris, 1672, Amsterdam, 1673. The Paris edition has considerable additional material.

Duhem, Pierre. *Congrès Scientifique International des Catholiques, compte rendu due troisième congrès,* i (1895), Bruxelles: Société Belge de Librairie, 313–15, 322–25.

———. 'Analyse de I'Ouvrage de Ernest Mach: La Mécanique, Étude Historique et Critique de son Développement', *Bulletin des Sciences Mathématiques,* i/xxvii (1903), 261–283.

———. 'Archimède a-t-ii Connu le Paradoxe Hydrostatique?, *Bibliotheca Mathematica* iii/i (1900), 15–19.

———. 'De l'Accélération Exercée par une Force Constante, Notes pour servir à l'histoire de la Dynamique', *Comptesrendus du Deuxième Congrès de Philosophie,* 1905, 859–915.

———. *De I'Aimantation par influence,* Paris, 1888. Includes thesis 'Sur I'Aimantation and Étude Historique from *Annales de la Faculté des Sciences de Toulouse* ii (1888), L2-L138, and do. ii (1888), 1–40.

———.*Die Wandlungen der Mechanik und der mechanischen Naturerklärung,* Autorisierte Übersetzung von Philipp Frank unter Mitwirkung von Emma Stiasny, Leipzig: J. A. Barth, 1912.

———.'Du Temps où la Scolastique Latine a connu la Physique d'Aristote', *Revue de Philosophie,* ix/i (1909), 163–178.

———. Étude sur l'Œuvre de George Green (à L'Occasion de la Réimpression des *Mathematical papers of the late George Green, Bulletin des Sciences Mathématiques*), ii/xxvii (1903), 237–256.

———. *Études sur Léonard de Vinci, ceux qu'il a lus et ceux qui l'ont lu,* 3 vols (vol iii subtitled *Les Précurseurs Parisiens de Galilée*), Paris, 1906, 1909, 1913.

———. 'François de Meyronnes OFM et la Question de la Rotation de la Terre', *Archivium Franciscanum Historicum,* vi/1 (1913), 23–25.

———. 'L'École Anglaise et les Théories Physiques, à propos d'un livre récent de W. Thomson', *Revue des Questions Scientifiques.* xxxiv (1893), 345–378.

———.*L'Évolution de la Mécanique*, Paris: A. Frienin, 1903. (from *Revue Générale des Sciences Pures et Appliquées*) (reprinted Paris: A. Hermann, 1905).

———. 'L'Évolution des Théories Physiques du xviie Siécle jusqu'à nos Jours', *Revue des Questions Scientifiques* xi (1896), 463–499.

———. *La Chimie, Est-Elle une Science Française?*, Paris 1916.

———. *La Science Allemande*, Paris, 1915. (includes 'Quelques Réflexions' on 103–143).

———. *La Teoria Fisica, il suo oggetto e la sua struttura, Collezione di Testi e di Studi, Filosofia*, Ed. Italiana a cura di S. Petruccioli. Traduzione Dario Ripa di Meana, Bologna: Il Mulino, 1978.

———. *La Théorie Physique, son Objet, sa Structure, Bibliothèque de Philosophie Expérimentale* (1st edn. Chevalier et Rivière, 1906; 2nd edn. 'revue et augmentée', 1914); *L'Histoire des Sciences, textes et études,* 'reproduction facsimile avec Avant-propos, Index et Bibliographie par Paul Brouzeng' (of 2nd edn., 3rd edn. Paris: Vrin 1951. First published in *Revue de Philosophie,* 1904–05.

———. 'Le P. Marin Mersenne et la Pesanteur de l'Air', *Revue Générale des Sciences,* xvii (15 and 30 September 1906), 769–782, 809–817.

———. *Le Système du Monde, histoire des doctrines cosmologiques de Platon à Copernic,* 10 vols, Paris: A. Hermann, 1913–59.

———. *Les Origines de la Statique, Les Sources des Théories Physiques,* 2 vols, Paris: A. Hermann, 1905, 1906. (*Revue des Questions Scientifiques.* 1903–06).

———. 'Les Théories de la Chaleur'. *Revue des Deux Mondes,* cxxix, cxxx (1895), 869–901, 380–415, 851–868.

———. *Les Théories Électriques de J. Clerk Maxwell, étude historique et critique,* Paris: A. Hermann, 1902. *Annales de la Société Scientifique de Bruxelles,* 2e partie (mémoires), xxiv, (1900), xxv (1901).

———. *Mediaeval Cosmology,* translation and selection by R. Ariew, Introduction by S. L. Jaki, Chicago: University of Chicago Press, 1985.

———. 'Notation Atomique et Hypothèses Atomistiques', *Revue des Questions Scientifiques* xxxi (1892), 391–454.

———. 'Notice sur les Travaux Scientifiques de Pierre Duhem', *Mémoires de la Société des Sciences Physiques et Naturelles de Bordeaux* vii/i (1917), 71–169.

———.'Physique de Croyant', *Annales de Philosophie Chrétienne* clv (1905–06), 44–67, 133–159 (*reprinted in Théorie Physique* 2nd edn., 413–447).

———. 'Physique et Métaphysique', *Revue des Questions Scientiques* xxxiv, (1893), 55–83.

———. 'Quelques Réflexions au Sujet des Théories Physiques', *Revue des Questions Scientifiques* xxxi (1892), 139–177.

———. 'Quelques Réflexions au Sujet De la Physique Expérimentale', *Revue des Questions Scientifiques* xxxvi (1894), 179–229.

———. 'Quelques Réflexions sur la Science Allemande', *Revue des deux Mondes* xxv (1915), 657–686. Reprinted in *Science allemande, 103–143.*

———. *Recherches sur l'Hydrodynamique,* 2 vols, Paris: Gauthier-Villars, 1903, 1904; *Publications Scientifiques et Techniques du Ministère de l'Air*, x, vi, iv, nouvelle édition complète dans un volume avec une préface de J. Kampé de Fériet, comprenanant aussi 7 notes aux *Comtes Rendus de l'Acadénie des Sciences* et la liste des publications de Duhem sur la mécanique des fluides, Paris, 1961.

———. *Salvar os Fenomenas, Cardenos de Historia e Filosofia da Ciencia, Supplemento*, Universidade Estadual de Campinas, Centro de Logica, Epistemologica et Historia da Ciencia, iii, 1984.

———. 'Science Allemande et Vertus Allemandes', in Petit, G., and Leudet, M., *Les Allemandes et la Science,* Cap xi, 1916, 137–152.

———. *ΣΩΖΕΙΝ ΤΑ φΑΙΝΟΜΕΝΑ, Essai sur la notion de théorie physique, de Platon à Galilée,* Paris: A. Hermann, 1908. *Annales de Philosophie Chrétienne* clvi, (1908), 113–138, 277–302, 352–377, 482–514, 576–592. Reprinted Vrin, Paris, 1982 (ed. Brouzeng).

———. *The Aim and Structure of Physical Theory,* translated by Philip P. Wiener, forward by Louis de Broglie, Princeton: Princeton University Press, 1954. Reprinted New York: Atheneum, 1964.

———. *The Evolution of Mechanics, Monographs, and Textbooks on Mechanics of Solids and Fluids,* introduction by G. Æ. Oravas, translated by Michael Cole, Alphen aan den Rijn, Germantown, Md: Sijthoff & Noordhoff, 1980.

———. 'Thermochimie, à propos d'un livre récent de M. Marcellin Berthelot'. *Revue des Questions Scientifiques* xlii (1897), 361–392.

———. *To Save the Phenomena, an Essay on the Idea of Physical Theory from Plato to Galileo,* translated from the French by Edmund Dolan and Chaninah Maschier, with an Introductory Essay by Stanley L. Jaki, Chicago: University of Chicago Press, 1969.

———. 'Un Précurseur Français de Copernic, Nicole Oresme', *Revue Génerale des Sciences* xx (15 November 1909).

———. 'Une Nouvelle Théorie du Monde Inorganique', *Revue des Questions Scientifiques* xxxiii (1893), 99–133.

———. *Ziel und Struktur der Physikalischen Theorien,* autorisierte Übersetzung von Friedrich Adler, mit einem Vorwort von Ernst Mach, mit einer Einleitung und Bibliographie herausgegeben von Lothar Schäfer, Hamburg: Felix Meiner Verlag, 1978. (Original edition J. A. Barth, Leipzig, 1912.)

———. *Le Mixte et la Combinaison Chimique, essai sur l'Évolution d'une Idée,* Paris: C. Noud, 1902. (from *Revue de Philosophie* 1901–02, reprinted Paris: Fayard, 1985).

Dühring, Eugen. *Kritische Geschichte der Aligemeinen Principien der Mechanik,* 1st edn. Berlin, 1873, 2nd edn. Leipzig, 1877.

Dupuy, Paul. *L'École Normale (1810–1883), Notice Historique, Liste des Eléves par Promotions, Travaux Littéraires et Scientifiques,* Paris, 1884.

Duthie, C. S. 'Pascal's Apology', *Congregational Quarterly* xxxiv (1956), 128–141.

Eastwood, D. M. *The Revival of Pascal, a Study of his Relation to Modern French Thought,* Oxford: Clarendon, 1936.

Favaro, Antonio. 'Galileo Galilei in una Rassegna del Pensiero Italiano nel Corso del Secolo Decimosesto', *Archivio di Storio della Scienza* ii (June 1921), 137–39.

Foucher, Simon. *Critique de la Récherche de la Vérité, où l'on examine en méme tems une partie des principes de Mr.*

Descartes, lettre par un academicien, Paris, 1675. Reprint edn. Watson, London and New York, Johnson Reprint Corporation, 1969.

Frank, Philipp. *Modern Science and its Philosophy,* Cambridge, Massachussetts: MIT Press, 1949.

Gardeil, A. *Revue Thomiste* (1894), 569–585, 738–759.

Geison, G. L. 'Pasteur, Louis', *Dictionary of Scientific Biography* x (1974), 350–416.

Gibbon, Edward. *The Decline and Fall of the Roman Empire,* 6 vols.

Gillispie, C. C. ed. *Dictionary of Scientific Biography,* 16 vols, New York: Scribners, 1970–1980.

Gilson, Étienne. *Christian Philosophy in the Middle Ages,* London, 1955.

———. *Le Thomisme, Introduction à la Philosophie de Saint Thomas d'Aquin, Études de Philosophie Mediéval,* Nouvelle édition, Paris, 1922, cinquième édition, Paris: Vrin, 1948.

Gossard, M. 'Sur les Frontières de la Métaphysique et des Sciences', *Revue de Philosophie* xii/i and ii (1912), 443–474, 575–586.

Gouhier, Henri. *Cartésianisme et Augustinisme au XVIIe siècle,* Paris: Vrin, 1978.

———. *La Jeunessee d'Auguste Comte et la Formation du Positivisme,* 3 vols, Paris, 1933–41. Vols ii and iii republished Paris: Vrin, 1964–70.

Grabmann, Martin. *Thomas Aquinas, his Personality and Thought,* Authorized translation by Virgil Michel, London, 1928.

Grant, Edward A. *Physical Science in the Middle Ages, Cambridge History of Science Series,* (previously Wiley, 1971), London: Cambridge University Press, 1978.

Grisar, Hartmann. *Galileistudien, Historisch-Theologische Untersuchungen über die Urtheile der Römischen Congregationen im Galileiprocess*, Regensburg: Fr. Pustet, 1882.

Hadamard, Jacques. *Archeion* xix (1937), 123–24.

———. 'L'Œuvre de Duhem dans son Aspect Mathématique', *Société des Sciences Physiques et Naturelles de Bordeaux, Mémoires* vii/ii (1927), 637–665.

———. *An Essay on the Psychology of Invention in the Mathematical Field*, translated by Jacqueline Hadamard from her father's French, Princeton: Princeton University Press, 1945.

Hall, A. R. 'Merton Revisited, or Science and Society in the Seventeenth Century', *History of Science* ii (1963), 1–16.

Hannequin, Arthur. *Essai Critique sur l'Hypothèse des atomes dans la Science Contemporaine*, Paris: G. Masson, Plon-Nourrot et Cie, 1895.

Hanotaux, Gabriel. *Histoire de la Nation Française*, 15 vols, Paris, 1920–35. Vol xiv *Mathématiques, Mécanique, Astronomie, Physique et Chimie*, by Henri Andoyer, Pierre Humbert, Charles Fabry, and Albert Colson.

Harding, Sandra G. *Can Theories be Refuted? Essays on the Duhem-Quine Thesis*, Dordrecht/Boston: D. Reidel, 1976.

Havet, Ernest. *Le Christianisme et ses Origines*, 4 vols, Paris: Michel-Lévy, 1871–84.

Hayek, F. A. *The Counter-Revolution of Science, Studies in the Abuse of Reason*, Glencoe, Illinois: Free Press of Glencoe, 1952.

Heath, T. L. *Aristarchus of Samos the Ancient Copernicus, a History of Greek Astronomy to Aristarchus together with Aristarchus's treatise on the sizes and distances of the sun and moon. A New Greek text with translation and gloss*, Oxford: Clarendon, 1913.

Hentschel, Klaus. 'Die Korrespondenz Duhem-Mach, zur "Modellbeladenheit" von Wissenschaftsgeschichte', *Annals of Science* xiv (1988), 73–91.

Hooykaas, Reye. *Religion and the Rise of Modern Science,* Edinburgh: Scottish Academic Press, 1972, 1973.

Hume, David. *Treatise of Human Nature*, ed. Selby-Bigge, Oxford: Oxford University Press, 1888.

Jaki, Stanley L. *The Physicist as Artist, the Landscapes of Pierre Duhem*, Selected and introduced by Stanley L. Jaki, Edinburgh: Scottish Academic Press, 1988.

———. *Uneasy Genius, The Life and work of Pierre Duhem. International Archives of the History of Ideas* c, The Hague: Martinus Nijhoff, 1984.

Jordan, Édouard. *Société des Sciences Physiques et Naturelles de Bordeaux, Mémoires* vii/i (1917), 3–40.

Knowles, David. *The Evolution of Mediaeval Thought*, London: Longman, 1962.

Koestler, Arthur. *The Sleepwalkers, a History of Man's Changing Vision of the Universe*, London: Penguin, 1964.

Krailsheimer, A. J. *Pascal, Past Masters,* London: Oxford University Press, 1970.

Kuhn, T. S. 'Notes on Lakatos', *Boston Studies in the Philosophy of Science* viii, Boston: Reidel, 1971, 137–46.

———. *The Essential Tension*, Chicago: University of Chicago Press, 1977.

———. *The Structure of Scientific Revolutions, International Encyclopedia of Unified Science* ii/ii, Chicago and London: University of Chicago Press, 1962, 1964.

La Rédaction [Maurice Blondel]. 'Notre Programme', *Annales de Philosophie Chrétienne* cii (1905–06), 5–31.

Laberthonnière, Lucien. *Esquisse d'une Philosophie Personnaliste*, ed. Louis Canet, Paris: Vrin, 1942.

———. *Le Réalisme Chrétien et l'Idéalisme Grec*, Paris, 1904.

———. 'Une Alliance avec Action Française', *Annales de Philosophie Chrétienne* cix (1910), 277–353.

Lafuma, Louis. *Histoire des Pensées de Pascal*, Paris: Éditions de Luxembourg, 1954.

Lagrange, J. L. *Mechanique analitique*, 1st edn., Paris, 1788; 2nd edn., 1811–15, 3rd edn., 1853–55.

Lakatos, Imre. 'Falsification and the Methodology of Scientific Research Programmes', *Criticism and the Growth of Knowledge*, 1970, 91–195. Reprinted in Imre Lakatos, *Philosophical Papers* i, ed. John Worrall and Gregory Currie, 8–101.

———. 'History of Science and its Rational Reconstructions', *Boston Studies in the Philosophy of Science*, ed. R. C. Buck and R. S. Cohen, viii, Boston: Reidel, 1971, 91–135. Republished in Lakatos, *Philosophical Papers* i, 102–138.

———. 'Proofs and Refutations, the Logic of Mathematical Discovery', *BJPS* xiv (1963–64). Republished ed. John Worrall and Élie Zahar, Cambridge: Cambridge University Press, 1976.

Lakatos, Imre, and Zahar, Élie G., 'Why did Copernicus's Research Programme Supersede Ptolemy's?, Lakatos, *The Methodology of Research Programmes; Philosophical Papers* i, (1978), 168-192. Originally in Westman (ed.) *The Copernican Achievement*, Los Angeles, 1976.

Lanson, Gustave. 'Pascal', *La Grande Encyclopédie* xxvi (1920), 20–30.

Lecanuet, E. *La Vie de l'Église sous Léon XIII*, Paris: Alcan, 1930.

Leibniz, Gottfried Wilhelm. *Discourse on Metaphysics and Related Writings, Classics of Philosophy and Science Series*, ed. R. N. D. Martin and Stuart Brown, Manchester: Manchester University Press, 1988.

———. *Die Philosophischen Schriften von G. W. Leibniz*, 7 vols, Berlin, 1875–90. Republished Hildesheim and New York, Olms, 1978.

Lemonnier, Henry. 'Les Études de Pierre Duhem sur Léonard de Vinci', *Journal des Savants, nouvelle série* xv (1917), 25–34, 120–132.

Lenin, V. I. *Materialism and Empiriocriticism, Critical comments on a Reactionary Philosophy*, Peking Foreign Languages Press, 1972. (Original edition 1909).

Leo XIII. 'Aeterni Patris', *Allocutiones, Epistolae, Constitutiones* i (1887). Translation (by Gilson) in J. Maritain, *St. Thomas Aquinas*, 183–214.

Libri, G. *Histoire des Sciences Mathématiques en Italie*, 4 vols, Paris, 1838–40.

Lowinger, Armand. *The Methodology of Pierre Duhem*, New York: Columbia University Press, 1941.

Mach, Ernst. *Die Mechanik in ihrer Entwicklung, Historisch-Kritisch dargestellt*, Leipzig, 1883.

———. *La Mécanique, Exposé historique et Critique de son Développement*, translated by Émile Bertrand, Professeur à l'École des Mines de Hainaut, intro. by E. Picard, Paris: A. Hermann, 1904.

Maiocchi, Roberto. *Chimica e Filosofia, Scienza, epistemologia, storia e religione nell' opera di Pierre Duhem, Pubblicazioni della Faccolta di lettere e filosofia dell' Università di Milano, Sezione a cura del Dipartimento di filosofia* cx/v, Firenze: La Nuova Italia Editrice, 1985.

Maire, Albert. *Bibliographie Généale des Œuvres de Blaise Pascal*, 5 vols, Paris: Girard-Badin, 1925–27.

———. *L'Œvre Scientifique de Blaise Pascal, Bibliographie critique et Analyse de Tous les Travaux qui s'y Rapportent,* Préface de Pierre Duhem, Paris: A. Hermann, 1912.

Malebranche, Nicolas. *De la Récherche de la Vérité ou l'on traite de la nature de l'esprit de l'homme, et de l'usage qu'il en doit faire pour éviter l'erreur dans les sciences*, 2 vols, Paris: André Pralard, 1674–78.

Mansion, Paul. *Congrès Scientifique International des Catholiques, Compte Rendu du Deuxième Congrès* viii (1891), 382–84.

Mansion, P., 'Note sure le Caractère Géométrique de l'Ancienne Astronomie', *Abhandiungen zur Geschichte der Mathématique,* ix (1899), 277–292.

Marchal, R., S. J. 'Symbolisme et Liberté dans les Sciences', *Revue de Philosophie* xi/i-ii (1911), 337–358, 489–510, 556–586.

Maritain, Jacques. *Distinguer Pour Unir, les Degrés du Savoir,* Paris: Desclée de Brouwer, 1932.

———. 'La Science Moderne et la Raison', *Revue de Philosophie* x/i (1910), 575–603.

———. *St. Thomas Aquinas*, New York: Meridian, 1958.

Martin, R. N. D. 'Darwin and Duhem', *History of Science* xx (1982), 64–74.

———. 'Saving Duhem and Galileo, Duhemian Methodology and the Saving of the Phenomena', *History of Science* xxv (1987), 301–319.

———. 'The Genesis of a Mediaeval Historian, Pierre Duhem and the Origins of Statics', *Annals of Science* xxxiii (1976), 119–129.

———. 'The Trouble with Authority, the Galileo affair and one of its Historians', *The Bulletin of Science, Technology, and Society* ix/v (1989), 294–301.

Martin, T. H. 'Mémoires sur l'Histoire des Hypothèses Astronomiques chez let Grecs et chez les Romains', Première partie, 'Hypothèses astronomiques des Grecs avant l'Époque Alexandrine', *Académie des Inscriptions et Belles Lettres, Mémoires* xxx/ii (1882), 21-foll.

McCool, Gerald A. *Catholic Theology in the Nineteenth Century, the search for a unitary method,* New York: Seabury, 1972.

Mentré, F. 'Pierre Duhem, le Théoricien', *Revue de Philosophie* xxix (1922), 449–473, 608–627.

Milhaud, Gaston. *Essai sur les Conditions et les Limites de la Certitude Logique*, 2e édition, Paris: Alcan, 1898.

Miller, Donald G. 'Duhem, Pierre Maurice Marie', *Dictionary of Scientific Biography* iv (1971), 225–233.

———. 'Ignored Intellect, Pierre Duhem', *Physics Today* xix/xii (1966), 47–53.

———. 'Pierre Duhem, un Oublié', *Revue des Questions Scientifiques* xxviii (1967), 445–470.

Montaigne, Michel de. *Œuvres Complètes*, ed. A. Thibaudet and M. Rat, Paris: Pleiade, 1962.

Montucla, J. E. *Histoire des Mathématiques*, 3 vols, 1st edn., Paris, 1758, 2nd edn., 1799.

Neurath, Otto. *Die Wissenschaftliche Weltauffassung, der Wiener Kreis*, herausgegeben vom Verein Ernst Mach, Vienna: W. A. Wolf, 1929. In Otto Neurath, *Empiricism and Sociology, the Vienna Circle Collection* i, Dordrecht: Reidel, 1973, 299–318.

Nye, Mary Jo. 'The Boutroux Circle and Poincaré's Conventionalism', *Journal of the History of Ideas* xi (1979), 107–120.

———. 'The Moral Freedom of Man and the Determinism of Nature: the Catholic Synthesis of Science and History in the

Revue des Questions Scientifiques', British Journal for the History of Science ix (1976), 274–292.

O'Rahilly, A. *Electromagnetic Theory, a Critical Examination of Fundamentals*, 2 vols, 1965. Original edition Cork University Press, 1938.

Oakley, Francis. *Natural Law, Conciliarism, and Consent in the Late Middle Ages, Studies in Ecclesiastical and Intellectual History,* London: Variorum Reprints, 1984.

Oregius, Augustinus. *De Deo Uno,* 1st edn., Roma, 1629, 2nd edn., 1630.

Owen, G. E. L. In *Aristotle, a collection of Critical essays, Modern Studies in Philosophy*, Penguin, 1968, 167–190.

Pascal, Blaise. *Œuvres Complètes*, présentation et notes de Louis Lafuma, Paris: Seuil, 1963.

———. *Pensées*, edited by Louis Lafuma, Paris, 1962: translated by Krailsheimer (Penguin 1966), and by Warrington (Everyman 1973).

———. *Pensées*, Translated with an introduction by A. J. Krailsheimer. Harmondsworth, Middlesex: Penguin, 1966.

———. *Pensées de Pascal, Publiées dans leur texte authentique avec un commentaire suivi et une étude litéraire*, by E. Havet, Paris: Dezobry et E. Magdaleine, 1852.

———. *Pensées*, ed. P. Sellier, Paris: Mercure de France, 1976.

Paul, Harry W. 'In Quest of Kerygma, Catholic Intellectual Life in 19th-Century France', *American Historical Review* lxxv (1969), 387–423.

———. 'Scholarship versus ideology, the chair of the General History of Science at the Collège de France, 1892–1913, *Isis* lxvii (1976), 376–397.

———. 'The Crucifix and the Crucible, Catholic Scientists in

the Third Republic', *Catholic Historical Review* lviii (1972), 195–219.

———. 'The Debate over the Bankruptcy of Science in 1895', *French Historical Studies* v (1968), 229–327.

———. *The Edge of Contingency, French Catholic Reaction to Scientific Change from Darwin to Duhem*, A University of Florida Book, Gainesville, Florida: University Presses of Florida, 1979.

———. *The Second Ralliement, The Rapprochement between Church and State in France in the Twentieth Century*, Washington, D.C.: Catholic University of America Press 1967.

Petit, G. and Leudet M. *Les Allemands et la Science*, Paris: Alcan, 1916.

Picard, Émile. *La Vie et l'Œuvre de Pierre Duhem*, Paris: Gauthier-Villars, 1922.

———. *Un Coup d'Œil sur l'Histoire des Sciences et des Théories Physiques*, Paris, 1930.

Picavet, François. *Esquisse d'une Histoire Générale et comparée des Philosophies Mediévales*, Paris: Alcan, 1905, 1907.

Pierre-Duhem, Hèléne. *Un Savant Français, Pierre Duhem,* Paris: Plon, 1936.

Pius X. 'Pascendi Dominici Gregis', *Actes de Pius X* iii (1908), 84-181. Translated as *Encyclical Letter ('Pascendi Gregis') of our most Holy Lord Pius X by Divine Providence Pope on the Doctrines of the Modernists,* London: Burns and Oates, 1907.

Popkin, Richard H. *History of Scepticism from Erasmus to Spinoza*, Berkeley and Los Angeles: University of California Press, 1979. Previous editions Van Gorcum 1960, Harper 1964.

Popper, K. R. *Conjectures and Refutations, The Growth of Scientific Knowledge*, 2nd edn. revised, London: Routledge, 1965.

———. *The Logic of Scientific Discovery*, London: Hutchinson, 1959.

———. *The Open Society and its Enemies*, 2 vols. 4th edn., London: Routledge, 1962.

———. *The Poverty of Historicism*, 2nd edn., London: Routledge, 1960.

Poulat, Émile. *Catholicisme, Démocratie, et Socialisme, le Mouvement catholique de Mgr. Benigni de la naissance du socialisme à la victoire du fascisme, Religion et Sociétés*, Paris: Tournai and Casterman, 1977.

———. *Histoire, Dogme, et Critique dans la Crise Moderniste*, 1st edn., Paris: Tournai and Casterman, 1962, 2nd edn., 1979.

Ptolemée, Claude. *Composition Mathématique*, ed. Halma, 3 vols, Paris, 1813–16.

Renan, Ernest. *Averroès et l'Averroïsme, essai historique*, Paris, 1852.

Rey, Abel. *Annales de Philosophie Chrétienne* cli (1906), 535–37.

———. La Philosophie Scientifique de M. Duhem', *Revue de Métaphysique et de Morale* xii (1904), 699–744.

———. *La Théorie Physique chez les Physiciens Contemporains*, Paris: Alcan, 1907. (Later editions substantially different.)

Rose, P. L. and Drake, S. 'The Pseudo-Aristotelian Questions of Mechanics in Renaissance Culture', *Studies in the Renaissance* xviii (1971), 65–104.

Rosen, Edward. *Three Copernican Treatises*, New York: Columbia University Press, 1939.

Rühlmann, M. *Vortäge über Theoretischen Maschineniehre*, Braunschweig, 1885.

Sarton, George and Tannery, Marie. 'Appel pour l'Achèvement du Système du Monde de Pierre Duhem', *Isis* xxvi (1937), 302-03.

Schiaparelli, G. V. *I Precursori di Copernico nell' Antiquità, recerche storiche, Reale Osservatorio di Brera, Publicazioni* iii, 1973.

———. 'Origine del Sistema eliocentrico presso i Greci', *Memorie del Instituto lombardo di Scienze e Lettere, classe de Scienze matematiche e naturali* xviii (1898).

Schmitt, Charles B. *Cicero Scepticus, International Archives of the History of ideas*, The Hague: Martinus Nijhoff, 1972.

Simmons, E. D. 'Sciences, Classification of', *New Catholic Encyclopedia* xii, 1222.

Simon, Walter M. *European Positivism in the Nineteenth Century, an Essay in Intellectual History*, Ithaca, New York: Cornell University Press, 1963.

Spain, B. 'Hadamard, Jacques', *Encyclopedia Judaica* vii, cols 1040-41.

Steinschneider, Moritz. 'Notice sur un Ouvrage Astronomique inédit d'Ibn Haïtam', *Bulletino di bibliografia e di Storia delle Scienze matematiche e fisiche pubblicato da B. Boncompagni* xiv (1883), 733–36.

Strowski, Fortunat. 'Le Secret de Pascal', *Le Correspondant* cxci (10 June 1923), 769–792.

Tannenbaum, E. R. *The Action Française: Die Hard Reactionaries in Twentieth-Century France*, New York, London: Wiley, 1972.

Tannery, Paul. *Mémoires Scientifiques* [correspondence with Duhem], eds. Heiberg and Zeuthen xiv, 207–226.

Testis [Maurice Blondel]. 'Les Moyens de"l'Univers"', *Annales de Philosophie Chrétienne* cx (1910), 346–353.

———. 'Méprises Révélatrices et Aveux Involontaires", *Annales de Philosophie Chrétienne* clx (1910), 252–276.

———. 'Correspondance, A. Cataine and Testis', *Annales de Philosophie Chrétienne* clx (1910), 187–89.

———. 'Dernières Réflexions sur le Système des "Alliances par les résultats seuls"', *Annales de Philosophie Chrétienne* clxi (1910), 263–285.

———. 'La Semaine Sociale de Bordeaux, Controverses dur les Méthodes et les Doctrines', *Annales de Philosophie Chrétienne* clix (1909–10), 162–184, 245–278, 372–392, 449–471, 561–592; clx (1910), 127–162.

———. 'Une Confirmation Imprévue de Mes Précédentes Critiques', *Annales de Philosophie Chrétienne* clx (1910), 69–78.

Thibault, Pierre. *Savoir et Pouvoir, Philosophie Thomiste et Politique Cléricale au XIXe Siècle*, Québec: Presses Universitaires de Laval, 1972.

Thirion Jules. *L'Évolution de l'Astronomie chez les Grecs, Revue des Questions Scientifiques*, Paris, 1899.

Venturi, G. V. *Essai sur les Ouvrages Physico-Mathématiques de Léonard de Vinci avec des Fragments Tirés de ses Manuscrits Apportés de L'Italie*, Paris, 1897.

Vicaire, Eugène. 'De la Valeur Objective des Théories Physiques', *Revue des Questions Scientifiques* xxxiii (1893), 451–510.

Vidler, Alec R. *A Variety of Catholic Modernists*, Cambridge: Cambridge University Press, 1970.

———. *The Modernist Movement in the Roman Church, its Origins and Outcome*, Norrisian Prize essay for the year 1933, London: Cambridge University Press, 1934.

Weber, Eugen. *Action Française, Royalism, and Reaction in Twentieth-Century France*, Stanford, California: Stanford University Press, 1962.

Weisheipl, James A. 'Neoscholasticism and Neothomism', *New Catholic Encyclopedia* x, 337.

Westfall, R. S. *The Construction of Modern Science, Mechanisms and Mechanics, Cambridge history of Science Series (Previously Wiley)*, Cambridge: Cambridge University Press, 1977 (previously New York, 1971).

Index

It has not been thought necessary to index 'Duhem', 'France', and 'Catholicism'. With the exception of encyclopedias and the like, works mentioned below by title and not credited to an author are by Duhem.